Springer Tracts in Modern Physics
Volume 123

W0232061

Springer Tracts in Modern Physics

---

* denotes a volume which contains a Classified Index starting from Volume 36

Particle Induced Electron Emission II

With Contributions by

D. Hasselkamp

H. Rothard, K.-O. Groeneveld, J. Kemmler

P. Varga, H. Winter

With 90 Figures

Springer-Verlag
Berlin Heidelberg GmbH

Dr. Dietmar Hasselkamp

I. Physikalisches Institut der Universität Giessen,
Heinrich-Buff-Ring 16, W-6300 Giessen, Fed. Rep. of Germany

Dr. Hermann Rothard
Professor Dr. Karl-Ontjes Groeneveld
Dr. Jürgen Kemmler

Institut für Kernphysik der Johann-Wolfgang-Goethe-Universität,
August-Euler-Strasse 6, W-6000 Frankfurt a. M. 90, Fed. Rep. of Germany

Dr. Peter Varga
Professor Dr. Hannspeter Winter

Technische Universität Wien, Institut für Allgemeine Physik,
Wiedner Hauptstrasse 8–10, A-1040 Wien, Austria

Manuscripts for publication should be addressed to:

Gerhard Höhler

Institut für Theoretische Kernphysik der Universität Karlsruhe, Postfach 69 80,
W-7500 Karlsruhe 1, Fed. Rep. of Germany

*Proofs and all correspondence concerning papers in the process of publication
should be addressed to*

Ernst A. Niekisch

Haubourdinstrasse 6, W-5170 Jülich 1, Fed. Rep. of Germany

ISBN 978-3-662-14981-2   ISBN 978-3-540-47447-0 (eBook)
DOI 10.1007/978-3-540-47447-0

© Springer-Verlag Berlin Heidelberg 1992
Originally published by Springer-Verlag Berlin Heidelberg New York in 1992.
Softcover reprint of the hardcover 1st edition 1992

Typesetting: Data Conversion by Springer-Verlag
Production Editor: P. Treiber

57/3140-5 4 3 2 1 0 – Printed on acid-free paper

Preface

This book deals with experimental aspects of electron emission from solid surfaces under bombardment by heavy particles in a wide energy range from eV to GeV. In three articles the state of the art in this field of physics is reviewed with the main emphasis on the basic aspects of the phenomenon. This volume provides a thorough overview which will help both the specialist on electron emission and physicists working in such fields as radiation effects in matter (e.g., radiation biology), particle induced emission phenomena, plasma-wall interactions, microtechnology, surface physics and surface analysis, atomic collision physics and others.

Electron emission by impact of energetic ions and neutral particles on polycrystalline and amorphous massive solids is summarized in the article by D. Hasselkamp, which includes a fairly complete tabulation of experimental work from the last twenty years. Electron emission from forward and backward surfaces of thin foils is treated by Rothard et al., with special emphasis laid on non-equilibrium phenomena. Varga and Winter discuss recent developments with respect to slow projectiles (particularly also highly charged), electron emission statistics and the interplay between potential and kinetic electron emission.

This book supplements Vol. 122 of this series, in which modern microscopic theories of electron emission are presented. Both books together provide a comprehensive account of the physics of 'particle induced electron emission'.

Giessen,   *D. Hasselkamp*
Frankfurt, and   *H. Rothard, K.-O. Groeneveld, and J. Kemmler*
Wien, September 1991   *P. Varga and H. Winter*

Contents

**Kinetic Electron Emission from Ion Penetration of Thin Foils
in Relation to the Pre-Equilibrium of Charge Distributions**
By *H. Rothard, K.O. Groeneveld* and *J. Kemmler*

Kinetic Electron Emission from Solid Surfaces Under Ion Bombardment

D. Hasselkamp

With 31 Figures

1. Introduction

Electron emission from solid surfaces under bombardment by energetic particles is a well-known emission phenomenon. Two different mechanisms must be distinguished. If the potential energy of the projectile is larger than twice the work function of the target material "potential electron emission" may proceed in front of the surface either by resonance neutralization followed by Auger deexcitation or by Auger neutralization (Hagstrum 1954, 1978). If electrons are excited within the solid by direct transfer of kinetic energy from the impinging projectiles the process is called "kinetic electron emission". It was first observed by Villard (1899) in experiments with canal rays and by Thomson (1904) and Rutherford (1905) for alpha particle bombardment of solid surfaces and was the subject of numerous review articles (Table 1.1). The present survey deals exclusively with kinetic electron emission from massive, polycrystalline or amorphous targets under heavy particle bombardment.

Two aspects served as the motivation for the continuous interest in this effect. On one hand kinetic electron emission is a basic phenomenon which accompanies all interactions of energetic particles with solid surfaces. It is intimately connected to the energy deposition in the target by the projectiles and is further strongly related to such fields as atomic collision physics, solid state physics and surface physics. On the other hand it is of general importance in applied physics as in plasma-surface interactions, in ion microscopies, in the development of heavy particle detectors and others. Renewed interest was motivated by the availability of UHV-techniques and surface analytical tools which allowed the performance of experiments on defined target surfaces under controlled conditions.

It is the aim of the present survey to summarize experimental work on kinetic electron emission in a comprehensive manner and to discuss the various aspects and facets of the phenomenon which emerged within the past twenty years. The review is intended to be a continuation of the work of Krebs (1968), i.e. experimental work before 1968 is only partly included. The point of view is that of an experimentalist, theoretical aspects are only covered briefly. The scope is restricted to heavy-particle induced kinetic electron emission from massive, polycrystalline or amorphous solids at impact energies of at least

Table 1.1. Review articles on ion induced kinetic electron emission

Author	Year	Topic
Rüchardt	1927	impact by canal rays
Geiger	1927	impact by α-rays
Massey and Burhop	1952	extensive survey
Little	1956	gas discharges
Petrov	1960	general considerations
Parilis	1962	theory
Möllenstedt and Lenz	1963	ion microscopy
Kaminsky	1965	extensive survey
Medved and Strausser	1965	extensive survey
Abroyan et al.	1967	short survey
Krebs	1968	extensive survey
Carter and Colligon	1968	extensive survey
Dearnaley	1969	short survey
Arifov	1969	extensive survey
McDonald	1970	short survey
Dettmann	1972	theory
Petrov	1974	special problems
Dorozhkin et al.	1974a	special problems
McCracken	1975	extensive survey
Krebs	1976	semiconductors and insulators
Krebs and Rogaschewski	1976	semiconductors and insulators
Baragiola et al.	1979b	short survey
Krebs	1980	special problems
Sigmund and Tougaard	1981	theory
Benazeth	1982	short survey
Krebs	1983b	short survey
Thomas	1984a	data for fusion
Ertl and Behrisch	1984	short survey
Hasselkamp	1985	extensive survey
Brusilovsky	1985	emission from single crystals
Lai et al.	1986	short survey
Benazeth	1987	Al-target
Hasselkamp	1988	short survey
Schou	1988	theory
Hofer	1989	extensive survey
Brusilovsky	1990	short survey

5 keV. Experiments dealing with emission from crystal targets or from thin foils are not considered, measurements of the statistics of electron emission are excluded from the discussion. Impact by multiply charged ions is only partly included at very high projectile energies.

2. Definitions and Basic Quantities

Several abbreviations have been used for the term "ion-induced electron emission": IIEE, IEE or IISEE (for ion-induced secondary electron emission). (It must be noted that the term "ion-induced electron emission" conventionally includes bombardment by neutral particles.) There is a philosophy which recommends not to use the term "secondary" with respect to ion bombardment in order to avoid confusion with "secondary electron emission" (SEE) which is for electron emission induced by primary electrons. In the following we will not use any of the above abbreviations. We will make use, however, of the term "secondary electrons" for those electrons which were excited in the course of the ion-solid interaction and which eventually may escape into the vacuum.

The term "δ-electrons" may need some clarification. Historically the name simply meant electrons which are emitted from a solid surface as a result of alpha-particle bombardment. The distinction between electrons ejected by canal rays and alpha-particles, respectively, was still made in the first reviews of the subject (Rüchardt 1927 for canal rays, and Geiger 1927 for alpha-particles). The name δ-electrons was later used by Bethe to denote ejected electrons with energy large compared to the ionization energy (e.g. Bethe and Ashkin 1960, p. 251). In this sense the term is used nowadays in the field of electron production in atomic collision physics. In the field of ion-induced electron emission the term was introduced again with a slightly different meaning by Sternglass (1957). Here δ-electrons denote energetic electrons produced in "close" collisions between projectile and target atoms.

From the viewpoint of an experimentalist who uses a special kind of accelerator it is a convenient practice to distinguish between low, medium and high projectile energies in a collision experiment. We will loosely define energy ranges as follows: "low" means below 25 keV, "medium" from 25 keV to 250 keV and "high" above 250 keV. This distinction obscures the fact that rather than the energy it is the projectile velocity that is the decisive quantity that governs the physical aspects of the interaction processes between projectiles and target atoms. Any comparison between results for different projectiles has to be performed at the same velocity (or the same energy per atomic mass unit).

In the following we will define the quantities that are of interest in an experiment on electron emission:

- The total electron yield γ is defined as the number of electrons per impinging primary particle which are emitted into the half space in front of the target (backward emission). The angle of incidence of the impacting projectile with respect to the surface normal of the target must be specified. Most experiments deal with normal incidence.
- The energy distribution of emitted electrons is either determined as the singly differential yield $d\gamma/dE$, i.e. integrated over all angles of emission,

or doubly differential with respect to both energy and angle as $d\gamma/dEd\Omega$. $d\gamma/dE$ as a function of the electron energy E is often referred to as the energy distribution curve (EDC) or simply the energy spectrum of emitted electrons. From its definition it follows: $\int (d\gamma/dE) \, dE = \gamma$. Instead of $d\gamma/dE$ also the notation $N(E)$ is in common use.

- The angular distribution $d\gamma/d\Omega$ of electrons outside the solid is usually only determined with respect to its dependence on the polar angle of emittance since for polycrystalline or amorphous solids no dependence on the azimuthal angle is expected.

At low impact energies potential emission may contribute to the measured electron yields γ if $E_i \geq 2e_0\Phi$ where E_i is the ionization energy of the projectile ion and $e_0\Phi$ the work function of the target material. Data for kinetic electron emission are usually corrected by substracting this contribution according to the calculation by Kishinevskii (1973)

$$\gamma_p = \frac{0.2}{E_F} \left(0.8 \, E_i - 2 \, e_0\Phi\right) \tag{2.1}$$

where E_F means the Fermi-energy of the target, or by an empirical relation obtained by Baragiola et al. (1979b) from a least-squares fit to experimental data on potential emission

$$\gamma_p = 0.032 \left(0.78 \, E_i - 2 \, e_0\Phi\right) \tag{2.2}$$

(energies in eV). For impact by singly charged ions at medium to high impact energies no correction is needed since it is generally acknowledged that potential emission yields decrease with increasing projectile energy (Hagstrum 1978) and is therefore negligible compared to the yields for kinetic emission.

3. Experiment

In this chapter we first summarize shortly the experimental conditions which should be met in studies of electron emission from solid surfaces (general environment, target preparation and characterization) and describe thereafter the measurement procedures for the total electron yield γ, the energy spectrum $d\gamma/dE$ or $d^2\gamma/dEd\Omega$ and the angular distribution of emitted electrons in some detail.

3.1 Experimental Conditions

Since electron emission stems from the first few layers of the sample the experimental environment should equal that of general analytical electron spectroscopies for surface analysis. In short the following conditions should be met in studies of electron emission: ultra-high-vacuum ($\approx 1 * 10^{-10}$ mbar), in-situ characterization and modification (cleaning) of the surface, avoidance of contamination of the surface during the measurement.

A mass-analyzed ion beam of specific energy and charge state must be used in the measurements. Isotope separation is generally not needed except at very low energies where elastic effects may be of some importance (Sect. 5.2.1). It should be noted that doubly charged molecular ions which are formed in the ion source can not be separated from singly charged ions of the same species in electrostatic or magnetic analyzers. Measurements taken with e.g. N^+ may therefore be influenced by an unknown fraction of N_2^{++} in the beam. Metastable particles in the beam will lead to an increased electron emission. The conditions within the ion source and electrical fields in the beam-line may be variied in order to check if changes of the electron yield occur. If this is not the case metastables may not be a serious problem.

Magnetic fields in the target chamber should be avoided. This is of special importance if the low-energy part of the electron spectrum is investigated (Sect. 3.3).

Some authors have performed the measurements on targets at elevated temperature in order to prevent adsorption of residual gas during the experiment or to maintain a well-defined surface structure during ion bombardment (Svensson and Holmén 1981). Elevated temperatures may also lead to annealing of defects near the surface and may be used to prevent charging of the target in the case of alkali-halide samples (König et al. 1975).

3.1.1 Target Preparation

The sensitivity of measurements to surface contamination was realized already early (around 1915). Much of the early work that was performed under moderate vacuum conditions is therefore only of qualitative value though general trends were correctly predicted. If comparison with other experiments is to be achieved or if the data are meant to be compared to theoretical predictions it is pertinent that the surfaces under investigation are characterized with respect to surface composition and probably topography. Unfortunately surface analytical tools (e.g. AES) were only seldom used in this context so far. In case that practical considerations prevail, e.g. for the study of charge-up under moderate vacuum conditions, data are needed if the surface is well characterized with respect to surface layers. Such studies are lacking, however.

The preparation of clean elemental solids was reviewed by Musket et al. (1982) and will not be repeated here. In-situ deposition of the target is preferable. In most cases the targets are mechanically polished or chemically etched in air and inserted into the vacuum chamber thereafter. In this case surface cleaning is necessary before starting the experiment. In electron emission studies cleaning is mostly achieved by sputtering either by the ion beam which is used for excitation itself or by an independent sputter-gun at low energies (≤ 5 keV). Problems related to sputter-cleaning have been discussed by Verhoeven (1978) and Taglauer (1990).

An alternative procedure of cleaning a surface from contaminants is the "filament-flash" method, i.e. the flash-heating of a metal strip to high temperatures (Krebs 1968 and references therein). The method is only applicable to materials with a high melting point ($\approx$ 2000 K). It has the disadvantage that increased diffusion and segregation of impurities from the bulk of the target may lead to an unwanted contamination of the surface (Ferrón et al. 1981b).

3.1.2 Target Characterization

Measurements of electron emission must be performed on well-characterized surfaces. Auger Electron Spectroscopy is especially suited as a surface-analytical tool since its information depth is comparable to the escape depth of secondary electrons. With respect to the achievement of clean metal surfaces a qualitative criterion is in common use. It is based on the observation that contaminated and/or oxidized elemental samples produce a higher electron yield than the clean metal surface. The following "rule of thumb" has been put forward: *A clean surface is achieved by sputter-cleaning with rare gas ions if the total electron yield γ has reached its lowest stable value* (e.g. Baragiola et al. 1979a, Alonso et al. 1979, Svensson and Holmén 1981, Benazeth 1982, Hasselkamp and Scharmann 1982b, Zalm and Beckers 1984). It was shown by Baragiola et al. (1979a,b) that the yield obtained after sputter-removal of an oxide film on pure aluminum by argon ion impact was the same as measured for an in-situ evaporated aluminum film (Fig. 3.1). If low-energy electron spectra are measured a further criterion was proposed: *The FWHM (full width at half maximum) of the true secondary electron peak is maximum for the clean surface* (e.g. Hasselkamp and Scharmann 1982b, Hasselkamp et al. 1987a). Examples for the application of these "rules" are given in Figs. 3.1,2. It must be noted that these methods of surface characterization are purely qualitative in nature and only based on experience. Furthermore they are not applicable

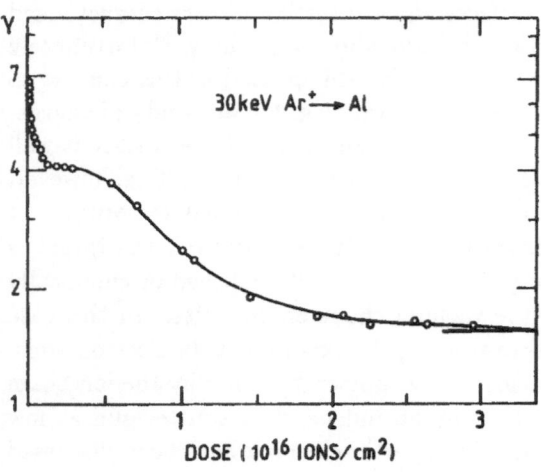

Fig. 3.1. Variation of the total electron yield γ as a function of the ion dose for an aluminum sample covered by an oxide film. The steady-state value (horizontal bar) at large doses corresponds to the yield from an in-situ evaporated aluminum film (from Baragiola et al. 1979b)

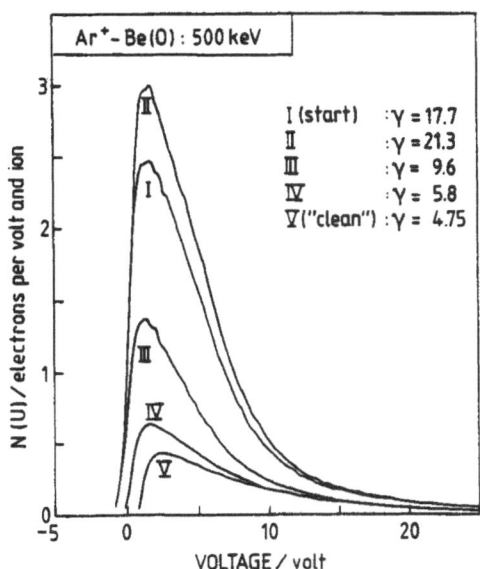

Fig. 3.2. Variation of the electron spectrum $N(U) = d\gamma/dU$ from a beryllium target covered by a thin oxide film during the course of sputter-cleaning by high-energy Ar⁺-ions (from Hasselkamp et al. 1987a)

to the investigation of electron emission from oxides and alloys where the use of AES is stringent.

There are two questions with respect to the surface characterization which so far have not been solved properly. The first is concerned with the modification of the target by implantation of ions during the experiments. No influence of implanted rare gas atoms has been found in studies by Baragiola et al. (1979a) and Zalm and Beckers (1984). The last authors remark that the equilibrium yield obtained under ion bombardment should be interpreted as a "steady-state" value rather than the yield for a virgin target. There is every reason to believe that implantation of reactive ions (e.g. oxygen) will influence electron emission. The result of implantation of hydrogen in niobium on the total electron yield gamma is demonstrated in Fig. 3.3. The yield decreases as a function of the ion dose with a corresponding increase of the work function of the surface. This is probably due to the formation of a hydride of niobium near the surface. Zalm and Beckers (1984) report on the variation of the yields from aluminum and copper during nitrogen ion bombardment which was attributed to the formation of a nitride layer at the surface. No systematic investigations of this interesting phenomenon are available so far. It must be concluded that low ion doses should be used in measurements. This requirement was not always fulfilled.

The second question is that of the influence of the surface topography. Again systematic investigation are sparse. Holmén et al. (1981) studied both the influence of a changing surface topography which developped during ion bombardment and the influence of the grain size of polycrystalline copper. Only slight changes of the electron yields were observed. Model calculations by Mischler et al. (1986) indicate that the yield from a textured surface is

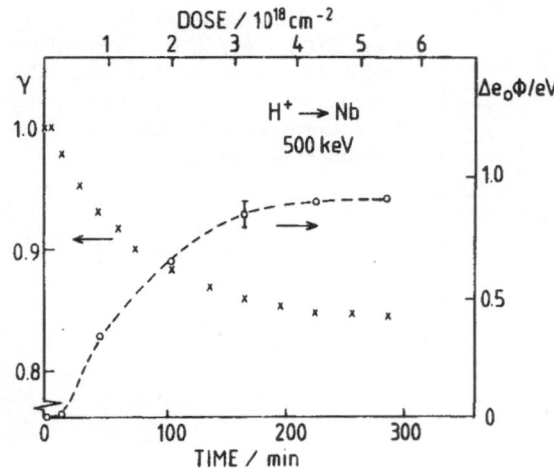

Fig. 3.3. The variation of the total electron yield γ and the work function $\Delta e_0 \Phi$ for sputter-cleaned niobium during proton bombardment (from Hasselkamp 1985)

higher than that from a flat surface (also Borovsky et al. 1988) while the angular distribution of emitted electrons remains unchanged (Sect. 5.5). Since erosion and modification of the target surface due to sputtering or implantation should be avoided as far as possible measurements of electron emission must be performed with low ion doses. This requirement was not always fulfilled.

3.2 Measurement of the Total Electron Yield

The determination of the total electon yield γ is straightforward for ion bombardment of metal targets. Two basically different methods are in use: i) the "quotient" method and ii) the ion-electron-converter method. Impact by neutral species or the bombardment of insulators require special care and will be treated only in short.

3.2.1 The Quotient Method

The quotient method directly refers to the definition of the total yield γ as the quotient of the number of emitted electrons to the number of incident primary particles. If charged primaries are used γ may be determined from the ratio of two currents or equivalently the ratio of two accumulated charges. Impact by neutral primaries will be discussed at the end of this section.

The measurement of γ for charged primary ions of (positive or negative) charge state z is easy in principle: for a given time T one determines the accumulated charge of the impinging ions Q_{ion} in a Faraday cup and afterwards the corresponding (negative) charge of electrons Q_e which are emitted from the target under consideration. Then γ follows from:

$$\gamma = z \frac{-Q_e}{Q_{ion}} \tag{3.1}$$

(here and in the following z and Q_{ion} are taken positive or negative depending on the charge of the ions while Q_e is always taken negative). Though extremely simple, this method is actually seldom used in routine measurements. The major disadvantages are:

i) a Faraday cup has to be inserted at or near the target;
ii) measurements have to be taken consecutively which may be problematic if ion currents are not stable or if fast changes of the yield have to be followed. Still the method is recommended at last for the purpose of calibration.

If a Faraday cup cannot be used the quotient method can be applied in two different ways: with and without an electron collector as discussed in the following.

Quotient Method Without Collector. In this case the target is biased alternately by a sufficiently high positive and negative potential (see Fig. 3.4). If a positive potential is applied with respect to the surroundings (i.e. the chamber walls) electrons are prohibited to leave the target and the accumulated charge Q_+ equals that of the incoming ion beam (Fig. 3.4a). In case of a negative potential electrons are repelled from the target (Fig. 3.4b) and the accumulated charge Q_- is the sum of the charges of incoming ions and outgoing electrons. For ions of charge state z the total electron yield follows as:

$$\gamma = z \left(\frac{Q_-}{Q_+} - 1 \right) . \tag{3.2}$$

Equation (3.2) holds for negative ion bombardment as well (z neg.).

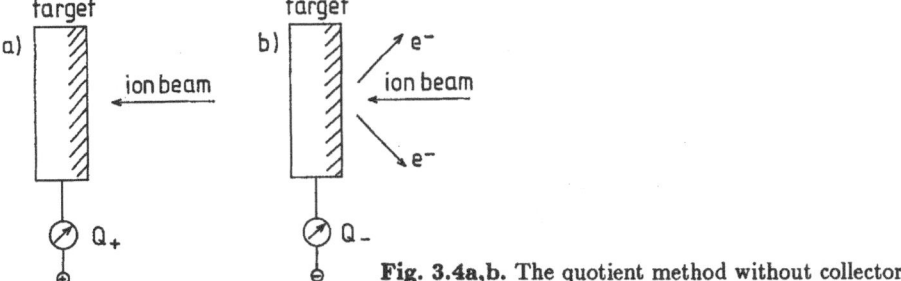

Fig. 3.4a,b. The quotient method without collector

Obviously this method shares one disadvantage with the one discussed above: measurements are taken consecutively. It is possible, however, to use a modulation method in this case (Krebs 1968, Arifov 1969). In recent years the modulation method was seldom used (e.g. Sørensen 1976, 1977, for insulating targets).

The maximum error associated with a measurement for positive ion impact according to (3.2) is in terms of the relative errors in Q_- and Q_+:

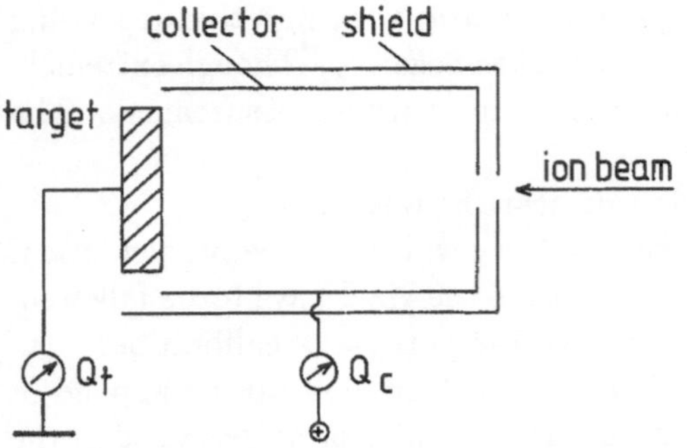

collector   shield

target

ion beam

Q_t Q_c

Fig. 3.5. The quotient method with collector

$$\frac{d\gamma}{\gamma} = \frac{z+\gamma}{\gamma}\left(\left|\frac{dQ_-}{Q_-}\right| + \left|\frac{dQ_+}{Q_+}\right|\right) \ . \tag{3.3}$$

The error will be large for $\gamma \ll 1$. Systematic errors will be discussed below.

The Quotient Method with Collector. A slightly more sophisticated set-up is in common use. Here the emitted electrons are collected by a separate collector electrode (cylinder or (half-)sphere) in front of the target (Fig. 3.5). Mostly the electron collector is biased positively while the target is kept on ground potential. In this case the accumulated charges resulting from currents from the target Q_t and from the collector Q_c can be measured simultaneously which is of importance if the ion beam is unstable or if fast changes of the electron yield are to be measured. In this case γ follows as:

$$\gamma = z\frac{-Q_c}{Q_t + Q_c} \tag{3.4}$$

(Q_c negative, $Q_{\mathrm{ion}} = Q_t + Q_c$). It should be noted that the target-collector assembly may be used as a Faraday cage. The maximum relative error for a determination of γ according to (3.4) is for positive ion impact:

$$\frac{d\gamma}{\gamma} = \frac{z+\gamma}{z}\left|\frac{dQ_t}{Q_t}\right| + \left|\frac{dQ_c}{Q_c}\right|\right) \ . \tag{3.5}$$

The relative error increases with γ which may lead to large errors in case of large γ (Decoste and Ripin 1979).

Errors Associatied with the Quotient Method. The quotient method assumes the existence of only two currents: the ion current and the current of emitted electrons. This is an idealized view since in reality for a surface under ion bombardment also other currents exist: sputtered secondary ions of both positive and negative charges; reflected ions; tertiary electrons from the surrounding walls (or from the collector) due to impinging electrons, reflected ions or sputtered particles; stray electrons e.g. from the defining slits of the ion beam. These currents constitute part of the systematic errors involved with the quotient method. Their relative importance depends on the ion-target-combination, the ion energy, the state of the surface and also the method used (either with or without collector):

Secondary Ions: Secondary ion emission depends on the value of the sputtering coefficient and on the probability for ion formation. For light ion bombardment the sputtering coefficient is mostly small compared to the electron yield. Therefore secondary ions will be of minor importance. For heavy ion bombardment, however, the sputtering yield may become comparable to the electron yield. Still for clean metal surfaces the probability for ion formation is low and secondary ions may be neglected (Wittmaack 1977, Blaise 1978). However, for contaminated – especially oxidized surfaces – and for insulators the secondary ion yield may become appreciable and comparable to the electron yield (Wittmaack 1977, Blaise 1978). In this cases the quotient method without proper correction fails to give meaningful results.

Reflected Ions: The reflexion coefficient for energetic ions is negligible for high energy bombardment but may be appreciable at low impact energies (Bøttiger et al. 1971, Sidenius and Lenskjaer 1976, Eckstein and Biersack 1983, Tabata et al. 1983).

Tertiary Electrons: All energetic particles which are emitted from the target due to ion bombardment may release tertiary electrons at the surrounding walls which in turn may strike the target. This is a problem especially for the quotient method without biased collector.

Stray Electrons: Stray electrons are formed at the apertures which define the primary ion beam and may reach the target or/and the collector. In general this can be easily prevented by proper bias potentials on the defining slits in front of the target chamber.

It follows from the above analysis that meaningful results for the measurement of electron yields are only accessible if either the disturbing currents are shown to be negligible or can be estimated or measured. This has not always been the case. While there are little problems (except for tertiary electrons) with clean metal surfaces, large currents due to e.g. secondary ions may be present in experiments on insulating surfaces. This (and the fact that insulators charge up under positive ion bombardment) is the reason why insulators were seldom investigated by the quotient method (see below).

Special care is required at high impact energies if electron loss from the projectile contributes to the measured electron yield (Sect. 5.2.4). While the quotient method with collector is still applicable care must be taken with the method without collector. The positive potential on the target must be high enough to draw all loss electrons back to the target. For an H_2^+-ion of 1 MeV the energy of the loss electron may amount up to 272 eV. Thus the potential on the target must be about +500 V in order to assure a valid measurement.

The determination of the total electron yield γ from insulators is complicated due to charging of the target during ion bombardment and due to possible erroneous current measurements by secondary ion emission. The charging of the target may be minimized i) by the use of very low ion currents, ii) by application of a pulsed beam method (Krebs 1968, Arifov 1969, Sørensen 1976), iii) for the alkali-halides by elevated temperatures (König et al. 1975). Charging may constitute no serious problem if thin insulating films on a conducting substrate are investigated. The quotient method is not applicable for extremely low ion currents. The use of the ion-electron-converter method is recommended (Sect. 3.2.2).

For the impact by neutral particles the quotient method is partly applicable for the measurement of the secondary electron current. However, the particle flux of neutrals must be determined seperately. Standard detection techniques for fast neutral particles have been summarized by e.g. Hasted (1964) and Krebs (1968). A much more elegant method for the measurement of the total electron yield γ can be achieved with an ion-electron-converter (see next section).

3.2.2 The Ion-Electron-Converter

The determination of the yield γ by the quotient method obscures the fact that on a microscopic scale electron emission is a stochastic process. A convenient method to study the emission statistics (i.e. the probability for the occurrence of 0,1,2,3,... emitted electrons per impinging primary particle) and the *mean* total electron yield is the ion-electron-converter (IEC) method.

In short the principles of operation are as follows. Electrons emitted from the target are dragged by a high potential (20–30 keV) onto a detector whose output is proportional to the deposited energy or equivalently to the number of electrons which arrive (almost) simultaneously. Mostly surface barrier detectors have been used. Different set-ups are possible (Krebs 1968, Hofer 1989, Varga and Winter in this volume).

Two standard probability functions have been proposed to describe the statistics of electron emission: while many authors have found the Poisson-distribution sufficient for a fit to their data others have reported that the two-parameter Polya-distribution yielded better results. The Poisson-distribution is of the form (for the single-event-γ: 1,2,3...):

$$P(\gamma, \overline{\gamma}) = \frac{\overline{\gamma}^{\gamma}}{\gamma!} e^{-\overline{\gamma}} \tag{3.6}$$

where $\overline{\gamma}$ is the average (or mean) electron yield; it has the meaning of a standard deviation for the Poisson distribution. The Polya-distribution can be written as

$$P(\gamma, \overline{\gamma}) = \frac{\overline{\gamma}^{\gamma}}{\gamma!} (1 + b\overline{\gamma})^{-\gamma - 1/b} * \prod_{i=1}^{\gamma} [1 + (i - 1) b] \tag{3.7}$$

which for $b = 0$ reduces to the the Poisson-distribution (Dietz 1970, Dietz and Sheffield 1973). The physical meaning of the parameter b is not clear. Dietz and Sheffield (1975) argued that the Polya-distribution may result from a superposition of Poisson-functions with fluctuating $\bar{\gamma}$. This may be induced by different conditions at the surface during the process of electron emission (e.g. differing topography at the point of electron emission). Still at present artefacts resulting from instrumental deficiencies cannot be ruled out. It was pointed out by several authors that the superiority of a two-parameter-fit is trivial and must not necessarily be taken as an argument against the basic Poisson-distribution (Hofer 1989, Lakits 1989). Lakits et al. (1989b,c) have argued that neither the Poisson- nor the Polya-distribution are valid fits to the measured pulse height spectra.

The mean electron yield $\bar{\gamma}$ may be derived directly by fitting the measured pulse-height spectrum by the proper probability function (3.6,7). The main advantages and draw-backs of the method can be summarized as follows (Hofer 1989):

Advantages

- very low ion currents are needed ($\leq 10^{-15}$A); surface damage and charging of the target are no problems;
- experiments are possible with particles which are only available with low beam currents, i.e. clusters, highly charged ions, metal ions;
- yield measurements with neutral particles are possible since no current measurement is involved.

Disadvantages

- the proper probability function describing electron emission must be known, the use of a Poisson- or a Polya-distribution was criticized (see above); in general the method is not applicable if the mean electron yield is low (≤ 1) since in this case the fit with a given probability function is not possible to any certainty;
- the method is especially sensitive to stray electrons or electrons arising from field emission;
- all electrons leaving the target must be collected by the solid state detector; this is inherently possible with the mirror-converter-device (Hofer 1989) but may be difficult otherwise; also backscattered electrons at the surface of the detector must be considered;
- there is no discrimination against negative ions from the target (suitable magnetic fields at the target or time-of-flight techniques may be used).

Typical errors are larger than with the quotient method, a comparison of results obtained by the two methods shows good agreement for $\gamma \geq 1$ (Ferguson and Hofer 1989, Lakits et al. 1989a, Lakits and Winter 1990).

3.3 Measurement of Energy Spectra of Emitted Electrons

The energy distribution curve of emitted electrons (in short the energy spectrum) may be measured singly differential, i.e. for all electrons emitted into the half space in front of the target, or doubly differential, i.e. differential both with respect to energy and the angle of emittance. The singly differential energy spectrum, denoted by $d\gamma/dE$, (or $N(E)$) gives the number of electrons of energy E per energy interval of 1 eV and per impinging primary ion which are emitted into all directions in front of the target. Instruments for the measurement of the low-energy part of the energy spectrum ($E \leq 500$ eV) are usually based on the principle of the retarding field analysis (RFA): a retarding field in front of the target allows the escape of only those electrons which have a kinetic energy exceeding $e_0 * U$ (e_0 the elementary charge, U the retarding potential).

Though in principle all set-ups capable of determining the total yield γ may also be used to measure the energy spectrum only the concentric sphere analyzer (CSA) and the concentric hemi-spherical analyzer (e.g. the LEED-system) have gained practical importance (Fig. 3.6). A short remark is necessary with respect to the simplest method, i.e. the set-up without a collector. A variable positive bias U at the target may also provide a retarding field and the yield $\gamma(U)$ may be determined by the method outlined in Sect. 3.2.1. In fact many authors have used this method. The main shortcomings of this procedure are i) the difficulty in establishing a symmetrical electric field in all directions in front of the target and ii) severe problems with stray electrons which are attracted towards the target. Thus this method only provides an estimate of the true energy distribution especially at low electron energies.

In the following we will discuss mainly the features of the CSA. This device was used by many authors (Krebs 1968 and, e.g., Losch 1970, Dorozhkin et al. 1976b, Gaworzewski et al. 1972/73, Hasselkamp et al. 1981, Hasselkamp and Scharmann 1982b).

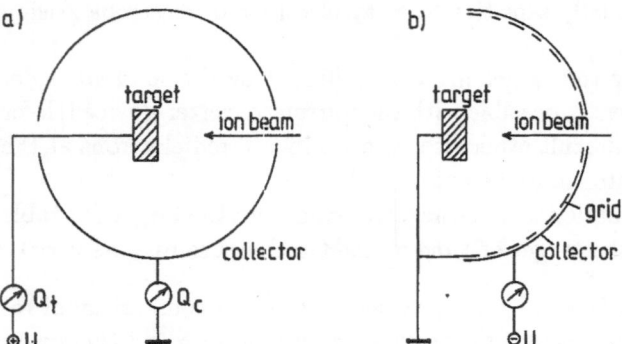

Fig. 3.6. The concentric sphere analyzer (a) and the concentric hemi-spherical analyzer with grids (b)

The CSA-device was first investigated by Lukirski (1924) and reexamined by Simpson (1961) and di Stefano and Pierce (1970). The energy resolution in the ideal case is given by (Simpson 1961)

$$\frac{\Delta E}{E} = \left(\frac{R_T}{R_C}\right)^2 \tag{3.8}$$

where R_T and R_C denote the radii of the target and collector sphere, respectively. E.g. by choosing $R_T = 10$ mm and $R_C = 80$ mm a resolution of 1.6 % can be achieved which is comparable to conventional LEED-systems (Hasselkamp and Scharmann 1982b). In case of a magnetic field of strength B inside the analyzer the energy resolution is degraded (di Stefano and Pierce 1970)

$$\Delta E \leq \frac{\left(e_0 B (R_C^2 - R_T^2) + R_T \sqrt{8 m_e E}\right)^2}{8 m_e R_C^2} . \tag{3.9}$$

It was pointed out by di Stefano and Pierce (1970) and by Gaworzewski et al. (1974) that an electron of energy E is measured in the interval $E - \Delta E$ and E. For a meaningful measurement of the energy distribution it is vital that the magnetic field strength is kept low ($\leq 5\mu$T). A suitable Mu-metal shield around the (non-magnetic) analyzer may serve this purpose.

The procedure for the determination of the energy spectrum is based on the quotient method (Sect. 3.2.1) and consists of three steps:

1) the electron yield is measured as a function of the retarding voltage U which yields $\gamma(U)$, the so-called integrated spectrum;
2) by differentiation of $\gamma(U)$ with respect to U the spectrum $d\gamma/dU$ is found as a function of the retarding voltage;
3) transformation of the voltage scale into an energy scale yields the energy spectrum $d\gamma/dE$.

Steps 1 and 2 may be performed together if a modulating voltage is superimposed upon the retarding voltage and phase-sensitive detection is used at the modulation frequency (e.g. Roy and Carette 1977).

In a practical application the integrated spectrum is determined as a function of the retarding voltage U from the accumulated charges on target and collector (which are measured simultaneously) with the help of (3.4) (step 1). A computer for the data collection allows to apply multi-scaling techniques and smoothing procedures. Differentiation of the integrated spectrum (step 2) is also best performed with the help of the computer and yields the spectrum as a function of the retarding voltage.

The transformation of the spectrum $d\gamma/dU$ into the desired energy spectrum $d\gamma/dE$ (step 3) needs some further considerations. On very general grounds one must expect that the energy spectrum goes to zero as the energy of emitted electrons tends to zero (e.g. Hachenberg und Brauer 1959, Schou

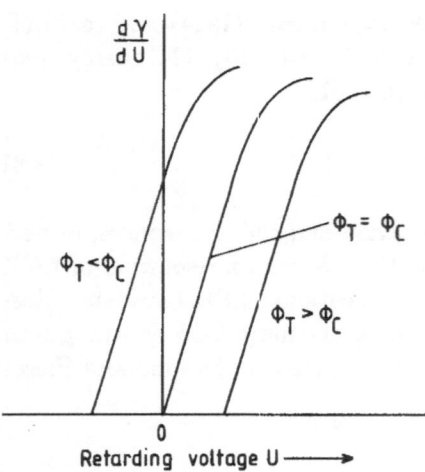

Fig. 3.7. The low-energy cut-off of the energy distribution curve $d\gamma/dU$ as a function of the retarding voltage U for different surface potentials of target and collector, respectively (positive retarding voltage applied to target)

1980, Sect. 4.1.3). In a practical analyzer with sufficient energy resolution the low-energy cut-off of the spectrum $d\gamma/dU$ is mostly not observed at zero retarding voltage: usually a small but definite offset is observed which is due to the difference in the work functions of target and collector or equivalently to the contact potential difference $\Delta\Phi = \Phi_T - \Phi_C$ (with Φ_T and Φ_C the surface potentials of target and collector, respectively). For a positive retarding voltage at the target we have the situation outlined in Fig. 3.7: If $\Phi_T < \Phi_C$ the cut-off occurs at a negative retarding voltage U, if $\Phi_T > \Phi_C$ at a positive U, only if the contact potential is zero the spectrum starts at $U = 0$. The transformation of the U-scale into an E-scale then simply means a shift of the energy distribution curve to start at zero or

$$E = e_0 U - e_0 \Delta\Phi \tag{3.10}$$

where E is the electron energy and e_0 the elementary charge. It shall be mentioned that a shift of the spectrum $d\gamma/dU$ may be used to measure a variation of the contact potential and eventually the change in the work function of the target (Palmberg 1967, Chang 1974, Hasselkamp et al. 1980, Benazeth et al. 1988b).

The energy calibration may be checked by comparing the energy position of well-known structures in $d\gamma/dE$ with data from the literature. We mention two such calibration points: 1) the main peak in the argon excited aluminum spectrum around 63.2 eV (Benazeth et al. 1977, Baragiola et al. 1982) and 2) the gold N_7VV-Auger-structure which is located at 69.8 eV (Powell et al. 1982a,b). Both structures are readily observed in ion-induced electron spectra above about 10 keV impact energy.

Additional to the degradation in energy resolution by a magnetic field (see above) the measurement of energy spectra is influenced mainly by the same error sources as already outlined for the quotient method in Sect. 3.2.1. The problem of tertiary electrons which may falsify both target and collector currents is, however, more severe than in the case of a measurement of the total electron yield γ. This effect will influence the low-energy part of the spectrum. A correction procedure is not known. The influence of tertiary electrons may be minimized by introducing a grounded grid in front of the collector and by applying the retarding potential at the collector while the target is kept at ground: since the space between target and grid is field-free the probability that tertiary electrons are not captured by the grid and hit the target is low. This device is now similar to the familiar LEED-Auger-systems (see below).

Despite its problems the concentric sphere analyzer without grids is an easy-to-operate system which has one outstanding advantage over other devices: since it is operated in the quotient method mode the energy distribution curves are evaluated on an absolute scale.

We now turn towards the other retarding field method which is the well-known LEED-Auger-system (or concentric hemi-spherical analyzer). Since its operation is well-documented in the literature (e.g. Roy and Carrette 1977, Rivière 1983, Seah 1989) its use for the measurement of low-energy electron spectra will be only touched upon in short. Though mostly used in investigations of Auger-spectra this device is equally well suited for the measurement of low-energy electron spectra. In principle the LEED-Auger-system consists of a hemispherical collector with three or four grids. It is usually operated by a modulated retarding field: by tuning the phase-sensitive detection system to the modulation frequency the energy spectrum $d\gamma/dE$ is measured, by tuning to twice the modulation frequency the derivative of the spectrum is detected. Both modes can be applied, in the second case an integration is necessary to obtain the energy spectrum.

In general the target current need not to be measured, the resulting energy spectra are evaluated on a relative scale. Relative spectra may, however, be made absolute by a normalization of the integrated spectra to the total yield measured in a separate experiment since $\int (d\gamma/dE)dE = \gamma$. While the energy resolution of LEED-systems and the concentric sphere analyzer are comparable the first does not suffer from problems with tertiary electrons from the collector (see above). The region between the target and the first grid must be free of electric and magnetic fields.

Standard electrostatic analyzers, e.g. the parallel plate analyzer, the hemi-spherical analyzer or the cylindrical mirror analyzer may be used to measure the doubly differential electron yield $d^2\gamma/dEd\Omega$. The experimental procedures are well documented (Roy and Carrette 1977, Rivière 1983, Seah 1989). An interesting alternative is the segmented concentric hemi-spherical analyzer which is described in the next section.

3.4 Measurement of the Angular Distribution

By the angular distribution is meant the differential electron yield $d\gamma/d\Omega$ measured as a function of the polar angle ϑ with respect to the surface normal (no dependence of $d\gamma/d\Omega$ on the azimuthal angle is expected for polycrystalline and amorphous targets). Two methods have been used:

1) An electron collector which can be rotated around the beam spot allows the measurement of the current of emitted electrons in the solid angle $\Delta\Omega$. Both a simple metal strip (Abbot and Berry 1959) or a Faraday cage (Klein 1965) served as an electron collector. In a more sophisticated device Mischler et al. (1984) used a Faraday cage behind the grids of a LEED system which allowed also to measure the doubly differential yield $d^2\gamma/dEd\Omega$ (Fig. 3.8).

2) A versatile instrument is the segmented concentric hemisperical (Soszka and Soszka 1980, Mischler et al. 1986) or spherical (Probst and Lüscher 1963) analyzer with grids (Fig. 3.8). It allows the measurement of the electron yield differential with respect to the angle of emittance ϑ and (or) with respect to the electron energy (Sect. 3.3). The set-up of Mischler et al. (1986) consisted of 27 concentrical segments which allows to achieve an angular resolution of $\Delta\vartheta \approx 3°$.

Since usually only a relative measurement is involved the primary ion beam must not be determined. It is necessary, however, that a stable ion beam is used during the measurement if a movable Faraday cage is used. In the case of the segmented analyzer the currents from all segments can be measured simultaneously. A major advantage of the segmented analyzer over the Faraday cage is the larger solid angle at a given ϑ because of the integration over the azimuthal angle.

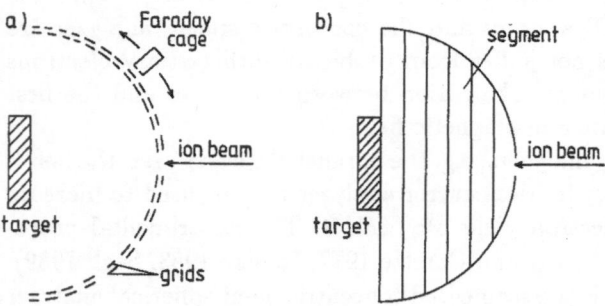

Fig. 3.8a,b. Measurement of the angular distribution. (a) rotatable Faraday cage behind LEED system; (b) segmented concentric hemispherical analyzer

4. Theory

Theoretical aspects of ion-induced electron emission will be discussed from the point of view of an experimentalist. In Sect. 4.1 we will investigate some general assumptions which are used in most treatments. In Sect. 4.2 the main results of various theoretical approximations are summarized as far as they lead to closed expression for the final results. No derivations will be given. Selected effects which so far have not been included in all theories will be studied in Sect. 4.3. We restrict ourselves mainly to metal and semiconductor targets and further to normal incidence of the incoming projectiles. The situation is less clear for insulators for which separate theoretical treatments are missing.

4.1 General Considerations

The mechanism of kinetic electron emission is considered to consist of three steps

- generation of excited electrons in the solid;
- diffusion of part of the excited electrons towards the surface including cascade multiplication;
- penetration of electrons through the surface into the vacuum.

It is a basic assumption of all theoretical treatments that the steps in this model may be treated separately. In the following we will give some general remarks for each of the mentioned steps.

4.1.1 Generation of Secondary Electrons

Electrons may be produced by a variety of processes, e.g.:

1) direct collision processes between projectile and target atoms and ions:
   a) by excitation of conduction (or valence) electrons into free states above the Fermi-level;
   b) by ionization of inner shells of the target atoms;
   c) by ionization in outer and inner shells of the projectiles;
   d) electron loss of electrons from the projectile;

2) by secondary processes:
   a) cascade multiplication of diffusing secondary electrons;
   b) excitation of target electrons by energetic recoil atoms (recoil ionization) and by backscattered projectiles;
   c) one-electron-decay of volume and surface plasmons generated either by energetic primary ions or by secondary electrons;
   d) by photons produced in projectile-target collisions.

It becomes clear immediately that a full treatment of the generation process alone is an immense task and in fact was only solved partly so far for the most important processes as there are: excitation and ionization of target electrons from the conduction band and from inner shells, plasmon decay (Sect. 4.3.2) and recoil ionization (Sect. 4.3.1). Other effects are of minor importance with the exception of the electron-loss process (Sect. 4.3.3). Electron production from the projectile was never explicitly included in theoretical treatments.

Many theories (in the following denoted as "semi-empirical") do not seperate between different excitation mechanisms; they rather treat the electron generation process in a semi-empirical way. Based on the observation that the total electron yield and the mean inelastic (or electronic) stopping power $(dE/dx)_e$ of the projectile follow the same dependence on the impact energy (Bethe 1941, Massey and Burhop 1952) the generation rate is assumed to be proportional to the ratio $(1/J)(dE/dx)_e$, where J is the mean energy to produce a free electron within the solid. This approach is reasonable as a first rough approximation since in $(dE/dx)_e$ all processes are included which contribute to the energy loss of a penetrating ion through inelastic (electronic) collisions and which consequently add to the energy transferred to electrons in the solid (including projectile ionization). An obvious deficiency is the fact that the generation process is related to mean quantities. It is assumed that the mean energy J is approximately independent of the impact energy and also of the ion-target combination (e.g. Bethe and Ashkin 1960, Boring et al. 1965). Without further assumptions a prediction of the energy- and angular-distribution of excited electrons is not easily possible. An advantage of this approach is the introduction of a well-known quantity, namely the inelastic stopping power, into the theory of ion-induced electron emission.

So far we have not yet discussed the question of the origin of secondary electrons. There is no serious problem at high impact energies: most authors assume that the main production mechanism is the excitation of conduction electrons from the Fermi-edge (Baragiola et al. 1979a,b, Sigmund and Tougaard 1981) and the ionization of inner-shell electrons (see also Rösler and Brauer 1984, 1988, 1991). The situation is more difficult, however, near the threshold of kinetic electron emission. It was pointed out by Ploch (1951) that direct energy transfer from a slow heavy ion to an electron at rest may not be sufficient to give the electron enough energy to overcome the surface barrier because of the unfavourable mass ratio. It was observed, however, that electron emission already proceeds at ion energies which were too low to account for the necessary energy transfer. Therefore an alternative excitation mechanism was proposed by Petrov and coworkers (e.g. Petrov 1960) and finally incorporated into the well-known theory of Parilis and Kishinevskii (1960). It was assumed that a heavy ion first produces an inner-shell hole in a target atom, subsequently an electron may be excited from the conduction band as a result of an Auger-effect. Later the conception of a direct excitation from the Fermi-sea was again discussed by Baragiola et al. (1979a,b) who pointed out that an electron in the conduction band is not at rest but moves with the Fermi-velocity v_F. In this case the maximum energy transfer T is much larger and by consideration of the conservation laws is given by

$$T = 2m_e v_p (v_p + v_F) \tag{4.1}$$

where m_e denotes the electron mass and v_p the velocitiy of the heavy projectile. Baragiola et al. (1979a,b) assumed that this energy transfer must at least equal the work function $e_0 \Phi$ for an emission event to occur. This leads to a threshold velocity v_{th} for kinetic electron emission by a direct ion-electron interaction of

$$v_{th} = 0.5 \, v_F \left(\sqrt{1 + e_0 \Phi / E_F} - 1 \right) \tag{4.2}$$

with E_F the Fermi-energy and Φ the surface potential. Baragiola et al. (1979a,b) and also Alonso et al. (1980) claim good agreement of their experimental results with (4.2) for proton impact. They noted that the experimental threshold was smaller than predicted by (4.2) for heavy ion bombardment. For this case these authors argued that another mechanism is working, namely excitation by quasi-molecular processes (e.g. Fano and Lichten 1965, Barat and Lichten 1972, Wille and Hippler 1986).

Recently expression (4.1) was reexamined by Hofer (1989) who concluded that it can hardly be expected to give the true threshold velocity, since it assumes a head-on collision which is very improbable and further that such a collision is forward-directed, i.e. into the solid and away from the surface. Hofer (1989) pointed out that for heavy ion impact near the threshold ($v_p \leq 10^7$ cm/s) quasi-molecular processes must be considered for the excitation process.

By anology with ion-atom-collisions the distribution function of internal electrons with energy E_i after the generation process will be forward peaked (Rudd et al. 1966, Rudd and Macek 1972) with an energy dependence of $\propto 1/E_i^m$ with $m \approx 2$–3 (Rudd et al. 1966, Rudd and Macek 1972, Toburen and Wilson 1972, Sigmund and Tougaard 1981).

4.1.2 The Diffusion Process

On their way towards the surface excited electrons will suffer from collisions with other electrons of the target. Thus the energy transferred to an electron in a primary event will be shared by a large number of other electrons and a cascade will develop. The assumption of a cascade is firmly grounded on the fact that the energy and angular distribution of electrons outside of the solid are almost independent of the ion energy and the ion-target combination. In the most sophisticated theories the diffusion process is described by mean free pathes for elastic and inelastic collisions of excited electrons (or by the low-energy stopping power for these electrons, see Schou 1980) while the cascade generation is treated as a transport theoretical problem (Schou 1980, Rösler and Brauer 1984, 1988, Dubus et al. 1986, Rösler and Brauer 1991, Devooght et al. 1991).

The semi-empirical theories on the other hand neglect the cascade process and consider only the diffusion process by the introduction of a mean free pathlength. The probability $f(x, E_i)$ for an electron of energy E_i inside the solid at a distant x below the surface to reach the surface is simply approximated as

$$f(x, E_i) \propto \exp\left(-\frac{x}{l(E_i) * \cos \vartheta}\right) \tag{4.3}$$

where ϑ is the direction of motion of electrons with respect to the (inner) surface normal. The mean free path $l(E_i)$ entering (4.3) must take account of both elastic and inelastic collision processes (Sigmund and Tougaard 1981). It is i) a function of the electron energy E_i and therefore depends on the energy spectrum of excited electrons and ii) is dependent on the target material (e.g. Tung et al. 1979, Penn 1987, Tanuma et al. 1990). Semi-empirical theories simply neglect the energy dependence of $l(E_i)$ and introduce a constant mean free path L.

At this point it is necessary to point out a tacit assumption of most theoretical treatments. They assume an isotropic distribution of excited electrons inside the solid. It was pointed out by Sigmund and Tougaard (1981) that this assumption is based on the experimental evidence that the angular distribution of electrons outside the solid closely follows a cosine law. On the other hand it is well known from experiment and theory of energetic ion-atom collisions that the original distribution will in most cases be anisotropic (Sect. 4.1.1). This problem was thoroughly discussed by Sigmund and Tougaard (1981) and Tougaard and Sigmund (1982). These authors suggest that many elastic collisions of excited electrons may be necessary for the final adjustment of an isotropical distribution. This view is strongly supported by the recent calculations of Rösler and Brauer (1984, 1988, 1991).

The mean free path of diffusing electrons may be used to give an estimate about the escape (or information) depth of secondary electrons which are detected in vacuum. Experimental mean free paths (e.g. Seah and Dench 1979) vary from about 5 Å to 20 Å for electron energies up to a few hundreds of eV for metals and semiconductors and may be considerable larger for insulating materials. Following the estimate of Seiler (1967) the information depth will be about five times the mean free pathes. Thus electron emission is a surface sensitive probe.

With respect to the energy spectrum it is generally assumed that its dependence $\propto 1/E_i^m$ (Sect. 4.1.1) is not altered by the diffusion process as derived by Wolff (1954), in a cascade treatment.

4.1.3 Penetration of the Surface Barrier

The surface is always approximated by a planar surface barrier characterized by the mean work function $e_0\Phi$ (with the surface potential Φ) and a barrier height $W = E_F + e_0\Phi$ in the Sommerfeld model (for semiconductors and insulators the electron affinity should be used instead of the work function, Schou 1980; the concept of a work function for insulators was questioned by Hofer 1989). Penetration of secondary electrons through the surface barrier is treated as a refraction phenomenon (Hachenberg and Brauer 1959, Schou 1980, Hofer 1989, Rösler and Brauer 1991).

An important consequence of this model is the fact that for a given energy E_i inside the solid only those electrons are able to surmount the barrier which have a velocity vector lying inside an escape cone with maximum angle ϑ^{max} with respect to the (inner) surface normal:

$$\vartheta^{max} = \arccos\left(\sqrt{W/E_i}\right); \qquad E_i \geq W \quad . \tag{4.4}$$

Assuming an isotropic distribution of secondary electrons inside the solid the probability $P(E)$ for the penetration of the surface barrier follows as

$$P(E) = 1 - W/E_i = (1 + W/E)^{-1}; \qquad E_i \geq W \tag{4.5}$$

where $E = E_i - W$ is the energy outside the solid (Chung and Everhart 1974, Alonso et al. 1979, Schou 1988).

Electrons with energies $E_i \leq W$ cannot overcome the surface barrier and will not be detected. For $E_i \geq W$ the escape probability is a function of the barrier height W: $P(E) \to 0$ for $E_i \to W$ and $P(E) \to 1$ for $E_i \gg W$ ($E \geq 0$). This fact has a direct consequence for the measured energy spectrum: for $E \to 0$ the spectrum should tend to zero in agreement with experimental evidence. Only for $E \gg W$ will the "outer" spectrum resemble the "inner" spectrum of secondary electrons. It is further evident from (4.5) that the escape probability decreases with increasing barrier height. Thus a change in the work function will also change the probability of emission.

If $N(E_i, \Omega_i)$ denotes the (isotropic) distribution function of secondary electrons then the flux density of particles of velocity v_i at the inner surface follows as

$$j_i(E_i, \Omega_i) = N(E_i, \Omega_i)\, v_i \cos(\vartheta_i) \tag{4.6}$$

giving rise to an angular distribution following a cosine law (ϑ_i with respect to the "inner" surface normal). It was pointed out by Jonkers (1957) that a planar surface barrier will not change the shape of the angular distribution of the particle flux. Thus it follows that also the outer flux density (corresponding to the outer angular distribution of secondary electrons) will follow a cosine law. A more detailed treatment may be found in Hachenberg and Brauer (1959), Schou (1980), Hofer (1989).

4.2 Summary of Main Theoretical Results

Within the last ten years sophisticated transport theories have become available (Rösler and Brauer 1984,1988, Dubus et al. 1986, Rösler and Brauer 1991, Devooght et al. 1991). They differ from the more phenomenological and semi-empirical theories by their detailed treatment of electron generation (including plasmon decay) and electron diffusion. However, they do not provide one with closed expressions i.e. for the total yield gamma, the energy distribution etc. and furthermore are restricted so far mostly to nearly-free electron metals like aluminum and to proton impact at fairly high energies. These important theories will not be considered below and the interested reader should consult the corresponding original papers and the reviews given in a preceding volume of this series (Rösler and Brauer 1991, Devooght et al. 1991).

In the following we will shortly summarize the main results of the semi-empirical and related theories. These approaches have been reviewed several times already (e.g. Medved and Strausser 1965, Krebs 1968, Carter and Colligon 1968, Arifov 1969, Baragiola et al. 1979b, Hasselkamp 1985). It is common to these treatments that distinct sources of electron emission are neglected like plasmon decay, electron loss, Auger-electrons, etc.. Furthermore only normal incidence ion impact on the solid surface is considered, for other angles of incidence ϑ the total yield is supposed to vary as $\sec \vartheta$ which simply accounts for the geometrical enhancement of the effective path length of ions within the escape depth of secondary electrons. An isotropic distribution of excited electrons inside the solid and a planar surface barrier are always assumed from which a cosine angular distribution outside the solid is a natural result.

The Semi-Empirical Theory. This approach is based on the so-called "elementary" theory of electron emission induced by electron bombardment and was first formulated by Salow (1940) (see also the reviews of Bruining 1954 and Dekker 1958). The theory was adapted by Baragiola et al. (1979a,b) for energetic ion bombardment under the assumption that the ion range is large compared to the mean escape depth of secondary electrons. These authors derived for the total electron yield:

$$\gamma = \frac{PL}{2J} * \left(\frac{dE}{dx}\right)_e \qquad (4.7)$$

(the symbols were introduced already in Sect. 4.1.1). The yield γ is predicted to be proportional to the inelastic stopping power of the projectile in the solid. Since the factor PL/J is considered to be a material parameter the energy dependence of the yield γ is governed by the stopping power. Thus for projectile energies below the maximum of the stopping power $\gamma(E) \propto \sqrt{E}$ and beyond the maximum $\gamma(E) \propto (1/E) \log E$ is expected from the theories of Firsov (1958, 1959) and Bethe (1930). Since the stopping power increases with the square of the effective charge Z_{eff} of the projectile this should also hold for the total electron yield:

$$\gamma \propto Z_{\text{eff}}^2 \quad . \tag{4.8}$$

In the derivation of (4.7) neither the inner angular nor the inner energy distribution of electrons is considered. Recoil effects are neglected which limits the applicability to ion impact energies where the inelastic stopping power is much larger than the elastic stopping power (see Sect. 4.3.1). The main weakness of this approach is the use of ill-defined quantities like the "mean" free path L, the "mean" escape probability P and the "mean" energy J needed to produce a secondary electron. Also the stopping power is a mean quantity which does not separate between different excitation mechanisms.

A more detailed treatment of the semi-empirical theory is possible (Hasselkamp 1985) from which an expression for the energy distribution may be derived as a function of the inner electron energy E_i:

$$d\gamma/dE = \frac{1}{4} * (1 - W/E_i) * L(E_i) * n(E_i) \tag{4.9}$$

where an isotropic distribution function was assumed and the penetration of the surface barrier was treated along the lines of Sect. 4.1.3. In (4.9) $n(E_i)$ is the energy distribution function (number of electrons with energy E_i per unit depth and per unit energy) and $L(E_i)$ is the mean free path for electrons of energy E_i. It has been suggested (Hasselkamp 1985) that as a first approximation the inelastic mean free path may be substituted for $L(E_i)$ (e.g. as given by Seah and Dench 1979) and that $n(E_i)$ may be approximated by an inverse power of the energy $\propto E_i^{-m}$ with an exponent $m = 2$-3 (see Sect. 4.1.1).

The Theory of Sternglass (1957). This extensive treatment is similar to the semi-empirical theory in many respects. The generation process is treated in more detail. Sternglass assumes that electrons are produced by two processes: 1) by so-called "distant" (or large impact parameter) collisions with a small energy transfer which give rise to a large number of low-energy secondary electrons; 2) by "close" (or small impact parameter) collisions with a large energy transfer which result in the excitation of a small number of energetic δ-electrons which may produce further excited electrons by cascade processes. The generation rates are set proportional to the corresponding inelastic stopping powers $(dE/dx)_d$ and $(dE/dx)_c * f(v_p, x)$ (the indices indicate distant and close, v_p is the ion velocity and x the depth below the surface). From the Bohr–Bethe equipartion rule of stopping powers (Bethe 1930, Bohr 1948) $(dE/dx)_d = (dE/dx)_c = (\frac{1}{2})(dE/dx)_e$ where $(dE/dx)_e$ denotes the total inelastic stopping power. The occurrence of the function $f(v_p, x)$ which enters together with the concept of close collisions must be explained since it has an interesting physical meaning. Sternglass argues that δ-electrons pro-

25

duced in close collisions will have directions which are pointed forward into the solid. Furthermore the mean free paths of these electrons will be larger than those of slow electrons. Sternglass concludes that energy transferred to the δ-electrons may be deposited and converted to slow secondary electrons at larger depths (compared to the original point of interaction) from where electrons may be unable to escape. The function $f(v_p, x)$ denotes that fraction of energy originally tranferred to δ-electrons which is transformed into slow electrons within the escape depth. According to the model this fraction will decrease with increasing impact energy. Sternglass gives an estimate of this function based on the different mean free paths of slow and of δ-electrons.

The total electron yield is finally derived as

$$\gamma = \frac{P'L}{2J} * \left(\frac{\mathrm{d}E}{\mathrm{d}x}\right)_e * (1 + F(v_p)) \qquad 0.5 \geq F > 0 \qquad (4.10)$$

where P' is an escape probability and the function $F(v_p)$ is related to the mean free paths of slow and of δ-electrons, respectively. $F(v_p)$ is a slowly decreasing function with increasing projectile velocity. We note the close similarity of this result with that of the semi-empirical theory.

Sternglass formulated his theory for high energy proton impact (> 100 keV) mainly because he used the Bohr and Bethe equations for the stopping power. If $(\mathrm{d}E/\mathrm{d}x)_e$ is available from other sources, e.g. from tabulations, (4.10) should also be valid for heavy projectiles and for lower impact energies as long as i) the energy loss is constant over the escape depth of electrons, ii) recoil effects are neglected and — very important — iii) the equipartition rule is fulfilled. It was pointed out by Baragiola et al. (1979a,b) that besides other points already raised above in the case of the semi- empirical theory the use of the equipartition rule may be criticized. As shown by Lindhardt and Winter (1964) equipartition is not fulfilled in general. An interesting approach to check the validity of the equipartition rule on the basis of the Sternglass theory was recently reported (Koschar et al. 1989 and Rothard et al. in this volume).

The theory is not applicable at low impact energies since the slowing down of projectiles within the escape depth is not considered. An extension of the Sternglass approach towards low impact energies was given by Beuhler and Friedman (1977a).

The Theory of Parilis and Kishinevskii (1960). This is the only theoretical approach which does not consider the excitation of conduction electrons as a possible source of excited electrons. Instead an ionization event in the valence band is anticipated followed by the excitation of electrons from the conduction band as the result of an Auger process. The generation of electrons was described by an effective ionization cross section σ^* (Firsov 1958, 1959), the expression for the Auger process ω (including the escape probability) was taken from the work of Hagstrum (1954). The total electron yield γ was then found as

$$\gamma = N * \sigma^* * L * \omega \tag{4.11}$$

where L is the mean free path of electrons and N the number density of the target. Equation (4.11) was predicted to hold for impact energies from threshold to about 100 keV.

The effective cross section σ^* was evalutated down to the threshold with inclusion of the stopping process of the projectile. Near the threshold a dependence of γ as a function of the impact energy E_p was predicted as $\gamma \propto E_p^2$, for higher energies as $\gamma \propto E_p$ and for projectile velocities larger than $3 * 10^5 \text{m/s}$ as $\gamma \propto \sqrt{E_p}$. The theory is the only one up to date which predicts a threshold of kinetic electron emission (around $0.5 - 0.6 * 10^5$ m/s).

An explicit dependence of γ on the atomic number of the projectile Z_1 (for a given target of atomic number Z_2) is predicted from the corresponding dependence of the effective cross section as

$$\gamma \propto \left(\frac{Z_1 + Z_2}{\sqrt{Z_1} + \sqrt{Z_2}} \right)^2 \tag{4.12}$$

for $1/4 \leq Z_2/Z_1 \leq 4$ and later on also for light ion impact as (Kishinevskii and Parilis 1962):

$$\gamma \propto \left(\sqrt{Z_1} + \sqrt{Z_2} \right) \left(\sqrt[6]{Z_1} + \sqrt[6]{Z_2} \right)^3 . \tag{4.13}$$

It must be noted that (4.12,13) do not give the Z_2- dependence of γ (as was wrongly assumed, e.g., by Krebs 1980), since according to (4.11) γ is also dependent on other factors than the effective cross section all of which are also dependent on target properties.

In the original version the theory neglected recoil ionization. This deficiency was later removed by Vinokurov et al. (1976).

Though the theory was very successful in numerous cases its concept gave rise to serious criticisms (Medved and Strausser 1965, Ferrón et al. 1981b). Discrepancies with experimental results were noted by e.g. Brusilovsky and Molchanov (1971), Krebs (1976), Krebs and Rogaschewski (1976). The following arguments were raised against the theory: i) wrong use of the Firsov-model; ii) total neglection of excitation of conduction electrons; iii) the ionization cross section does not match the energy dependence of γ; iv) wrong prediction of Z_1-dependence. It was recently noted by Brusilovsky (1990) that some of the criticized points may be irrelevant.

The Theory of Schou (1980). This approach is based on the ionization cascade theory, results are obtained from the solutions of a system of Boltzmann transport equations. Main input quantities are the cross sections for collisions between interacting particles, which are approximated by general power cross sections, and the magnitude of the plane surface barrier. The treatment partly includes energy transport by recoiling electrons and backscattered primary particles and is able also to consider recoil ionization. A phenomenological survey of this theory was given by Sigmund and Tougaard (1981) and by Schou (1988).

If recoil ionization is neglected the electron yield is derived as a remarkable simple formula:

$$\gamma = \Lambda * D_e \tag{4.14}$$

where D_e is the amount of inelastic energy deposited at the surface and Λ is given as

$$\Lambda = \int_0^\infty \frac{\Gamma_m E \, dE}{4 \, |dE_i/dx| \, (E + W)^2} \tag{4.15}$$

with $E = E_i - W$, Γ_m a function dependent on the exponent of the used power cross section (tabulated by Schou 1980) and dE_i/dx the stopping power for low-energy electrons in the solid.

We note that the expression for the total yield is given as a product of two terms, one of which contains the dependence on the characteristics of the impacting ion (the quantity D_e), while the other depends only on properties of the target material and is therefore considered to be a material constant (Λ).

Equation (4.14) may be written to contain the inelastic stopping power of the impacting ion $(dE/dx)_e$:

$$\gamma = \Lambda * \beta * \left(\frac{dE}{dx}\right)_e \tag{4.16}$$

where the factor β accounts for energy transport by recoiling electrons (similar to the concept of δ-electrons in the theory of Sternglass 1957) and by backscattered ions, it is a slowly varying function of the impact energy. For proton impact on Be, Mg and Al at 100 keV $\beta \approx 0.3$ (Schou 1988).

While there is a striking similarity of (4.16) with the result of the semi-empirical theory (4.7) there are also distinct differences. Instead of the inelastic stopping power the deposited energy at the surface is used and instead of the mean free path the continuous slowing down of electrons is described by the low-energy stopping power dE_i/dx (first introduced by Bethe 1941). The inclusion of these quantities must be considered an important improvement over the simple semi-empirical theory (Sigmund and Tougaard 1981).

Schou gives an expression for the energy spectrum:

$$\frac{d\gamma}{dE} = \frac{\Gamma_m D_e E}{4 \, dE_i/dx \, (E + W)^2} \cdot \tag{4.17}$$

It is predicted that the shape of the energy spectrum depends on properties of the target but neither on the energy nor the type of the projectile. This is a general characteristic of a cascade process.

If ionization by energetic recoil atoms is included the total yield is given by

$$\gamma = \Lambda * (D_e + D_r) \tag{4.18}$$

where D_r denotes that portion of deposited energy at the surface which stems from recoil ionization events. It is proportional to the elastic (nuclear) stopping power $(dE/dx)_n$ of the projectile of energy E_p in the target material:

$$D_r = \beta_r \left(\frac{dE}{dx}\right)_n \frac{\eta_n(\mu E_p)}{\mu * E_p} \quad . \tag{4.19}$$

μ is the maximum energy transfer factor $4M_p M_t / (M_p + M_t)^2$ (M_p and M_t denote the masses of projectile and target atoms, respectively). The factor $\eta_n(\mu E_p)$ accounts for the transfer of elastic deposited energy into electronic excitation. It is a slowly varying function of E_p (Schou 1980). Finally energy transport processes are included by means of the factor β_r which depends weakly on the impact energy but strongly on the mass ratio M_p/M_t (Schou 1980). A full transport theoretical treatment of D_r was given by Holmén et al. (1979), an approximate calculation which neglects energy transport (i.e. by recoiling electrons and by backscattered ions) was published by Holmén et al. (1981).

It was noted by Sigmund and Tougaard (1981) and by Schou (1988) that the cascade theory is valid only if a sufficiently large number of energetic electrons is excited in the primary interaction between projectile and target atoms. For low impact energies this assumption will not be true and the cascade treatment is not applicable. Schou (1988) has considered the case in which all secondary electrons stem from primary ionization. In this case the resulting expression for the total yield gamma becomes more complicated. Schou (1988) has shown that the yield should be proportional to the ionization cross section.

The sophisticated treatment of Schou (1980) is superior to older theories in several respects: it treats electron emission as a transport theoretical problem on the basis of a cascade process, still it gives closed expressions for the final results; all input quantities are properly defined. There are, however, several problems remaining: the range of validity of the cascade treatment is not clear, Schou (1988) estimates a lower limit of several keV for the impact energy; the inelastic stopping power for low-energy electrons is not known very well; the deposited energy must be calculated by a transport theoretical treatment if energy transport processes are to be included.

4.3 Special Effects in Electron Emission

There are several effects — which contribute to the total yield gamma on one hand or which lead to distinct structures in the energy spectra on the other — which are only partly included in existing theories or are neglected at all. Sizable too large contributions to the total electron yield may result from recoil ionization, plasmon decay, molecular effects and loss electrons, structures in the energy spectra arise from Auger-electron emission, loss electrons and plasmon decay.

4.3.1 Recoil Ionization

Elastic collisions between projectile and target atoms may lead to a sizable fraction of energetic recoil atoms which may produce excited electrons by themselves. This effect is called recoil ionization. First estimates of this additional production mechanism were given by Dorozhkin et al. (1974a,b) and Vinokurov et al. (1976). Detailed treatments were published by Holmén et al. (1979), Schou (1980) and Holmén et al. (1981). The effect was also considered by Alonso et al. (1980), Ferrón et al. (1981a,b) Svensson et al. (1981), Nagy et al. (1983), Giber et al. (1984).

It is immediately clear that elastic collisions will play no role as long as the elastic stopping power is small compared to the inelastic stopping power. A simple and rough estimate of the influence of elastic collisions was given by Sigmund and Tougaard (1981) in the energy range of the maximum of the elastic stopping power. They defined a parameter ξ which estimates the ratio of electronic excitation by elastic and inelastic collisions, respectively:

$$\xi = \frac{2}{3} \left(\frac{Z_t}{Z_p} \right)^{1/6} \left(\frac{Z}{Z_t} \right)^{2/3} \left(\frac{M_p}{M_p + M_t} \right)^2 \tag{4.20}$$

with

$$Z^{2/3} = Z_p^{2/3} + Z_t^{2/3} \tag{4.21}$$

where $Z_{p,t}$ and $M_{p,t}$ are the atomic numbers and masses of projectile and target atoms, respectively. It follows from (4.20) in the energy range of the maximum of the elastic stopping power: recoil ionization may be neglected for $Z_p \ll Z_t$, it contributes by $\approx 25\%$ for $Z_p \approx Z_t$, and is the main process for $Z_p \gg Z_t$.

The influence of recoil ionization on the shape of the energy spectrum was not investigated so far.

4.3.2 One-Electron Plasmon Decay

The interaction of fast charged particles with the electron gas of the target may lead to the excitation of plasma oscillations (Raether 1980). This process is related to electron emission in several respects: i) it contributes to the energy loss of the projectile; ii) it influences the diffusion of secondary electrons as described by a mean free path or by the low-energy stopping power; iii) the decay of plasmons may lead to a transfer of the plasmon energy to an electron (one-electron plasmon decay).

Two different plasmon energies must be distinguished. Plasma oscillations in the volume of the target are characterized by the volume plasmon of energy $\hbar\omega_p$, those in the surface by the surface plasmon of energy $\hbar\omega_s$. For an interface metal-vacuum the relation between the energies of volume and surface plasmons is $\omega_s = \omega_p/\sqrt{2}$ (Raether 1980). If one-electron plasmon decay occurs and dispersion is neglected secondary electrons of energies

$$
\begin{aligned}
E_v &= \hbar\omega_p - e_0\Phi \\
E_s &= \hbar\omega_s - e_0\Phi
\end{aligned}
\tag{4.22}
$$

are expected to be emitted after penetration of the surface barrier of height $e_0\Phi$. Since energy degradation due to inelastic collisions during the diffusion to the surface will occur these electrons are expected to produce a contribution to the total energy spectrum which will extend from zero energy to a maximum energy given by (4.22) and which is therefore expected to give rise to shoulders in the energy spectrum around the energies E_v and E_s.

In the following we will only consider the decay of volume plasmons for the case of a nearly-free electron gas. Here ω_p is the long-wavelength (classical) plasma frequency

$$
\omega_p^2 = \frac{ne_0^2}{\varepsilon_0 m_e}
\tag{4.23}
$$

where n is the number density of free electrons, e_0 the elementary charge, m_e the electron mass and ε_0 the dielectric constant. The dispersion relation reads (Raether 1980):

$$
\omega^2(k) \approx \omega_p^2 + \frac{3}{5}v_F^2 k^2
\tag{4.24}
$$

where ω_p and k denote the frequency (4.23) and the wave vector of the volume plasmon and v_F the Fermi-velocity. Equation (4.24) holds for $0 \leq k \leq k_c$ where k_c is the so-called cut-off wave vector which in a first approximation is given by (Raether 1980):

$$
k_c \approx \omega_p/v_F \ .
\tag{4.25}
$$

With respect to electron emission the threshold velocity v^{th} is of interest below which no excitation of plasmons should occur. It follows from the conservation of energy and momentum of the (heavy) projectile and the plasmon, respectively, and with the dispersion relation (4.24) (Hasselkamp 1985, the treatment is an extension of the work of Everhart et al. 1976 for electron impact excitation):

$$\frac{v^{th}}{v_F} \approx \frac{k_F}{2k_c} \frac{\hbar\omega(k_c)}{E_F} \qquad (4.26)$$

(E_F and k_F are the Fermi-energy and the Fermi-wave-vector, respectively). We note that v^{th} is independent of the mass of the (heavy) projectile. For aluminum the theoretical value of $k_c = 1.3 * 10^{-8}\,cm^{-1}$ (Raether 1980) and $v^{th} = 2.27 * 10^8\,cm/s$ which leads to threshold energies of about 30 keV for proton impact and 1200 keV for argon ion impact. The proton value may be compared to the results of detailed calculations of about 40 keV (Erginsoy 1967) and of 35–40 keV (Rösler and Brauer 1984, 1988).

Electron emission from plasmon decay was calculated by a number of authors mainly for primary electron impact (e.g. Chung and Everhart 1977, Rösler and Brauer 1981, cf. also the review of Bindi et al. 1987) but also for proton impact (Rösler and Brauer 1984, 1988, 1991).

It must be noted that theories which are based on the inelastic stopping power take into account the energy loss of the projectile due to plasmon excitation but neglect the one-electron plasmon decay.

4.3.3 Electron Loss from the Projectile

Loosely bound electrons of the projectile may be scattered from the first layer of the solid surface in a quasi-elastic process known as electron loss from the projectile in ion-atom collisions. (e.g. Burch et al. 1973, Stolterfoht 1978, Rudd et al. 1980). If the projectile velocity v_p is large compared to the velocity of the bound electron, the electron velocity v_e after the scattering event is

$$v_e \approx v_p \ . \qquad (4.27)$$

A "loss" peak will be observed in the energy spectrum at an energy corresponding to the projectile velocity. The width of the "loss" peak can be estimated (Stolterfoht 1978):

$$\Delta E \approx 4\sqrt{E_e I_b} \ . \qquad (4.28)$$

where $E_e = 1/2 m_e v_p^2$ and I_b is the binding energy of the electron in the projectile before scattering. ΔE increases both with increasing projectile velocity and with increasing binding energy.

The theory of electron loss for ion-atom collisions was first formulated by Drepper and Briggs (1976). These authors showed that the cross section for this effect is strongly forward-directed.

With respect to electron emission in the backward direction the electron loss process may contribute i) by reflexion of loosely bound projectile electrons from the solid surface which increases the total electron yield and gives rise to a structure in the energy spectrum at the characteristic energy corresponding to (4.27) and ii) by generation of additional secondary electrons in the solid. Present theories of ion-induced electron emission neglect these processes.

4.3.4 The Molecular Effect

Electron emission induced by fast molecular ions cannot be understood simply in terms of an additive effect of the ionic constituents at the same projectile velocity. This observation is called "molecular effect". We will confine the following discussion to molecular hydrogen ions for which simple models have been developped. It is convenient to introduce a ratio (e.g. Rothard et al. 1990b)

$$R(v_p) = \frac{\gamma \, (\text{molecule}, v_p)}{\sum \gamma \, (\text{proton}, v_p)} \tag{4.29}$$

where v_p denotes the projectile velocity. Experimental evidence for molecular hydrogen ions is $R < 1$ at low to medium and $R > 1$ at high impact velocities (Sect. 5.2.4). Two approaches have been put forward which both assume that the total electron yield is proportional to the square of the effective charge of the projectile (4.8) and further that the interaction may be understood in terms of "close" and "distant" collision events.

In the model of Svensson and Holmén (1982) it is assumed that a molecular hydrogen ion H_n^+ ($n = 2, 3$) may be treated as

$$H_n^+ = H^+ + (n - 1)H^0 \tag{4.30}$$

where H^+ denotes a proton and H^0 a neutral hydrogen atom. These authors assume that for velocites below the threshold for plasmon excitation $\gamma(H^0) < \gamma(H^+)$ because of the lower effective charge of a H^0 compared to the proton which will lead to a "negative" molecular effect ($R < 1$). At high velocities these authors assume that the electrons of the molecular ion are stripped in the first layers of the solid. Then in close collisions the molecular ion will act as the sum of the constituent protons. However, in distant collisions which lead to the excitation of plasma oscillations a proton cluster is considered which has a larger effective charge than the sum of independent protons due to the "vicinage" effect (Brandt et al. 1974, Brandt and Ritchie 1976):

$$\left(\sum_n Z_n\right)^2 > \sum_n (Z_n)^2 \, . \tag{4.31}$$

Since the yield γ is assumed to be proportional to the deposited energy and therefore to the square of the effective charge (4.8) a "positive" molecular effect is expected ($R > 1$). The projectile velocity must be high enough that excitation of plasmons is possible and the screening length in the electron gas must be larger than the separation of protons in the cluster (Svensson and Holmén 1982).

The above considerations finally lead to the formulation of a scaling relation which relates the yields γ of protons and molecular ions as

$$\gamma(H_3^+) = 2 * \gamma(H_2^+) - \gamma(H^+) + \Delta^* \tag{4.32}$$

where Δ describes the enhancement of the molecular ion yields as a result of the "vicinage" effect.

A slightly alternative approach was used by Hasselkamp et al. (1984, 1987b). These authors discriminate between distant and close collisions, respectively:

$$\begin{aligned} H_n^+ &= H^+ + (n-1)\,H^0 & \text{"distant"} \\ H_n^+ &= n\,H^+ + (n-1)\,e^- & \text{"close"} \end{aligned} \tag{4.33}$$

where "H^0" denotes a correlated proton-electron system and "e^-" a free electron, both at the projectile velocity. In "distant" collisions screening of the proton by the correlated electron is assumed which leads to a lower effective charge and a lower electron yield compared to the bare proton. At high projectile velocities the correlation between proton and electron will be destroyed by electron loss processes (stripping). In "close" collisions two situations are encountered: i) if the projectile velocity is low the co-travelling electron cannot contribute to electron emission; ii) at high projectile velocities the "extra" electrons give rise to a contribution by true secondary electron emission (see also Gosh and Khare 1962, 1963). Thus the net effect of both "distant" and "close" collisions is a "negative" molecular effect at low ($R < 1$) but a "positive" effect at high projectile velocities ($R > 1$). A similar model was used and discussed by Lakits et al. (1989a) for the explanation of the molecular effect and the differences in the electron yields by protons and neutral hydrogen atoms at low impact energies.

From both (4.33) a scaling relation may be deduced (Hasselkamp et al. 1987b)

$$\gamma(H_3^+) = 2 * \gamma(H_2^+) - \gamma(H^+) \tag{4.34}$$

which is the same as (4.32) in the model of Svensson and Holmén (1982) except for the absence of the "vicinage" correction term. It was shown by Hasselkamp et al. (1987b) that (4.34) may also be applied to the energy spectra induced by proton and molecular ions at the same impact velocity and is even valid for excitation and ionization cross sections in ion-atom collisions at high impact velocities.

We note for completeness that at projectile energies close to the threshold for kinetic electron emission a molecular effect may arise from the different potential emission of protons and molecular hydrogen ions (Dorozhkin et al. 1974a, 1975, Baragiola et al. 1978).

4.3.5 Auger–Electron Emission

Auger-electrons show up in the ion-induced energy spectra either as broad shoulders or in special cases also as sharp atomic-like peaks which are super-imposed on the background of the so-called "true" secondary electrons. This subject was extensively studied by many authors and has been summarized in several review articles (e.g. Hennequin and de Viaris 1980, Baragiola 1982, Thomas 1984b, Brusilovsky 1990). In the present survey we will touch upon this topic only very shortly.

Present theoretical treatments of electron emission do not include the emission of Auger-electrons. In fact Auger-electrons where shown to con-tribute only to some percent to the total yield γ at normal incidence (e.g. Benazeth et al. 1988a), still these electrons give rise to marked features in the energy spectra. The equivalence of proton and electron impact with respect to the resulting Auger-spectra was demonstrated by Musket and Bauer (1972) and Musket (1975). Differences between ion and electron bombardment were discussed by Matthew (1983). The influence of the diffusion process on the shape of Auger-structures was investigated experimentally by Toburen et al. (1982) and theoretically by, e.g., Sigmund and Tougaard (1981) and Tougaard and Sigmund (1982).

4.3.6 The Binary Encounter Peak

The maximum energy transfer of a fast heavy projectile of mass M_p and energy E_p to a quasifree electron (mass m_e) at rest is given by

$$E^{BEA} \approx \frac{4m_e}{M_p} E_p \cos^2 \theta \tag{4.35}$$

where θ ($0 \leq \theta < 90°$) is the scattering angle of the electron with respect to the direction of the incoming projectile. Around E^{BEA} a structure is found in the energy spectrum which is referred to as the binary-encounter-peak (Gerjuoy 1966, Vriens 1969, Garcia 1970). It follows from the kinematics of this process that the BEA-electrons are forward-directed. This effect is therefore of minor importance in measurements of "backward" electron emission except for those cases where non-normal or even glancing incidence of the ion beam is studied. We note, however, that for high energy impact E^{BEA} has the meaning of a cut-off energy for the energy spectrum of secondary electrons.

5. Summary of Typical Results

It is the purpose of the forthcoming chapter i) to summarize experimental work on ion-induced electron emission in a tabular form covering the period from 1968 to the beginning of 1990 and ii) to present and discuss selected results with main emphasis laid on clean or well-defined surfaces.

5.1 Summary of Experimental Work in the Period from 1968 to 1990

Tables 5.1–5.3 contain a fairly complete account of relevant experimental work published in the field of ion-induced kinetic electron emission in the period from 1968 until the beginning of 1990. Each table entry contains the following informations: the energy range of the experiment, the projectile-target combinations, the authors and the year of publication. The entries are arranged in the order of increasing maximum projectile energy.

The tables are meant to carry on the corresponding summary of Krebs (1968) which extends up to 1967 (some references have been added in the present tables which have been missing in the review by Krebs 1968). The contents are restricted to measurements with energetic projectiles (energy at least 5 keV) on massive targets. Experiments with crystalline targets and on the statistics of electron emission are partly included since they may provide general information which is not available otherwise. Also experiments are considered which deal mostly with practical aspects. Though there is a general trend to perform experiments under UHV-conditions within the last 15 years one must be aware of the fact that many studies quoted in Tables 5.1–5.3 have been undertaken under moderate vacuum conditions and on undefined surfaces. Such data are seldom of basic interest to the understanding of electron emission but may be important for practical reasons.

5.2 Energy Dependence of Total Yields from Metal Targets at Normal Incidence

This section deals with the energy dependence of total electron yields measured at normal incidence. We distinguish between impact by positive ions, neutral and negative projectiles, molecular ions and clusters. A special section is devoted to the relation between the total electron yields and the inelastic stopping powers of positive ions in the solid target. Only experiments on clean metal targets are considered.

Table 5.1. Experiments with projectile energies up to 25 keV; "many" indicates more than 10 different projectiles, number not specified in original paper; "(sol)" denotes solid gas target, SS a stainless steel target

E (keV)	Ions	Targets	Author	Year
5	He$^+$	Cu	Soszka	1973b
5	many	CuBe	Pottie et al.	1973
5	He$^+$	Ni	Soszka and Stepien	1981
5	H$^+$,H$_2^+$,H$_3^+$	Ni	von Gemmingen	1982
5	He$^+$,Ar$^+$	NiFe	Soszka and Soszka	1983b
5	He$^+$	Mo	Soszka and Soszka	1983c
0.1–5	H$^+$,H$_2^+$,H$_3^+$	W,Al,Mo	Losch	1970
0.12–5	H$^+$	Cu,Nb,Mo,SS,Cs	Ray and Barnett	1971
1–5	He^0,Ar^0	Al	Collins and Stroud	1967
1–5	He^0,He$^+$,He*	Au	Layton	1973
1–5	He$^+$,Ar$^+$	AgMg	Dev	1982
1–5	He$^+$,Ar$^+$,Kr$^+$,Xe$^+$,U$^+$	U	Chen et al.	1983
5.1	24 ions	Cu–Be	Lao et al.	1972/73
5.4	H(H$_2$O)$_x^+$,H(NH$_3$)$_x^+$,cluster	CuV	Nikolaev et al.	1978
6	He$^+$,C$_2$H$_2^+$,O$^+$	Au	Soszka and Soszka	1977
6	He$^+$,O$_2^+$	Si	Soszka	1978
6	He$^+$	Ni	Soszka and Soszka	1983a
0.2–6	He$^+$,Ne$^+$,Ar$^+$,Kr$^+$ He^0,Ne^0,Ar^0,Kr^0	KBr,NaCl,LiF	Arifov et al.	1971b
0.6–6	K$^+$	Mo,W	Breunig	1972
1–6	K$^+$	AgMg	Faizan-Ul-Haq and Chaudhry	1976
8	many	CuBe, Al, Ni	Fehn	1976
2–8	Xe$^+$	W	Bonnano et al.	1990
3–9	He$^+$,Ne$^+$	Cu	Soszka and Soszka	1980
10	He$^+$	Si, SiO$_2$	Dorozhkin et al.	1976b
10	He$^+$,Ar$^+$	Si	Dorozhkin et al.	1976a
10	NH$_3^+$,NP$_3^+$,many	CuBe	Märk	1977
10	He$^+$	Cu	Soszka et al.	1980
10	Ar$^+$,Kr$^+$,Xe$^+$	Cu,Au,Kr(sol),Xe(sol)	Soszka	1990
0.3–10	Xe$^+$	Au	Alonso et al.	1986
1–10	H$^+$,H$_2^+$,H$_3^+$,D$_2$H$^+$, D$_3^+$,He$^+$	H$_2$(sol), D$_2$(sol)	Børgesen et al.	1980
2–10	many	CuBe	van Gorkum and Glick	1970
2–10	H$^+$,H$_2^+$,H$_3^+$	HD(sol), H$_2$/D$_2$(sol)	Sørensen et al.	1983
3–10	H$^+$,H$_2^+$,H$_3^+$	H$_2$(sol), D$_2$(sol)	Sørensen	1976
4–10	H$^+$,H$_2^+$,H$_3^+$,D$_3^+$	H$_2$(sol), D$_2$(sol)	Sørensen	1977
6–10	Xe$^+$	FeNi,Xe(sol)	Soszka et al.	1989
15	many,cluster	SS	Staudenmaier et al.	1976
0.5–15	many	viele	Dorozhkin et al.	1974b
1–15	He$^+$,Ar$^+$	Mo	Wehner	1966
1–15	H$^+$,D$^+$,C$^+$,N$^+$,O$^+$,Ne$^+$, Ar$^+$,H$_2^+$	C,Ni,Ta,Pt	Cawthron	1971

Table 5.1. (continued)

E (keV)	Ions	Targets	Author	Year
1–15	H_2^+, Hg^+	C,Si,Ge,Ni,Fe,Cu,Nb, Re,Ta,Mo,W	Dorozhkin and Petrov	1974b
16	Ne^{2+}, Ar^{1+-4+}	Au	Lakits et al.	1989d
0.5–16	H_2^+, Hg^+	W,Mo,Ta,Nb,Cu	Dorozhkin and Petrov	1974a
1–16	$H^+, H^0, H^-, He^+, He^0,$ Ne^+, Ne^0, Ar^+, Ar^0	Au	Lakits et al.	1990
1–16	$H^+, D^+, H^0, H^-, H_2^+, H_3^+$	Au	Lakits et al.	1989a
1–16	H^+, H_2^+, H_3^+	Au	Lakits et al.	1989b
1–16	H^+, He^+, Ne^+, Ar^+	Au	Lakits et al.	1989c
1–16	$H^+, H^0, H_2^+, H_3^+, He^{0-2+},$ Ne^{0-3+}, Ar^{0-4+}	Au	Lakits and Winter	1990
1–17	$Ne^+, Kr^+, Ar^+, C^+, N^+, N_2^+$ $CO^+, CO_2^+, O^+, O_2^+, H_2O^+$	W,Mo,V	Cook and Burtt	1975
1–18	$Ar^+, CO^+,$ many	Cu	Dorozhkin et al.	1975
20	many	Au,Ag,Pd,Cu,Ti	Ferguson and Hofer	1989
1–20	Ar^+, Ar^0	Be,CuBe,Cu,Al	Schackert	1966
1–20	$H^0, H_2^0, He^0, N^0,$ N_2^0, Ar^0	Pt,W,Al,	Devienne	1967
5–20	$H^+, H_2^+, He^+, Ne^+, Ar^+,$ Kr^+, Xe^+	Al,Ti,Ni,Cu,Zn, Mo,Ag,Au,Pb	Zalm and Beckers	1984
5–20	$H^+, H_2^+, He^+, Ne^+, Ar^+,$ Kr^+, Xe^+	Cu,Zn	Zalm and Beckers	1985
21	$H^+, Li^+, Na^+, K^+, Rb^+, Cs^+,$ cluster	Al_2O_3, CsI	Moshammer and Matthäus	1989
0.3–22.4	$H^+, C^{1-6+}, Al^{1-10+}$ Cu^{1-9+}, Ta^{1-3+}	Au,Mo,CuBe	Cano	1973
6–24	He^+, Ar^+	Mo	Perdrix et al.	1970
1.5–25	He^+, Ne^+, Ar^+, Kr^+	Si	Gaworzewski et al.	1972/73
1.5–25	$He^+, Ne^+, Ar^+, Kr^+, Xe^+$	KBr,KCl,NaCl,LiF	König et al.	1975
2–25	Ne^+, Ar^+, Kr^+	Si,Ge,GaAs,GaP	Gaworzewski et al.	1974
12.5–25	$V_x^+, Nb_x^+,$ cluster	SS	Thum and Hofer	1979
12.5–25	$V_x^+, Nb_x^+, V_xNb_y^+,$ cluster	SS	Thum and Hofer	1980
12.5–25	$V_x^+,$ cluster	SS	Hofer	1980

Table 5.2. Experiments with projectile energies up to 250 keV; for special notations see Table 5.1.

E (keV)	Ions	Targets	Author	Year
26	many	SS	Rogaschewski and Düsterhoff	1976
29.6	Li^+, Rb^+	$Al_2O_3, CuBe$	Dietz and Sheffield	1973
30	Ne^+, Ar^+, Kr^+	Cu	Evdokimov et al.	1967b
30	Ar^+	Cu	Evdokimov and Molchanov	1968
30	Ar^+	Cu	Brusilovsky and Molchanov	1974
30	Sb^+, Sb_2^+, Sb_3^+	Ag	Veje	1981
30	Ar^+	Cu,Mo	Brusilovsky et al.	1985
1–30	He^+, Ne^+, Ar^+, Kr^+	NaCl	Baboux et al.	1971
2–30	many	Au	Thum and Hofer	1984
5–30	He^+, Ne^+, Ar^+, Kr^+	Mo	Perdrix et al.	1968
5–30	$Cl^+, Cl^-, O^+, O^-, H^+, H^-$	Mo	Perdrix et al.	1969a
5–30	H^+, H_2^+	Pt	Cawthron et al.	1970
5–30	$Li^+, Na^+, K^+, Rb^+, Cs^+$	$Al_2O_3, BeO, MgO, Ta_2O_5$	Dietz and Sheffield	1975
25–30	$He^+, N^+, Ne^+, Ar^+, Kr^+$	Cu,Ag,Mo,Zr,W,Bi	Evdokimov et al.	1967a
8-32	He^+	W	Dev	1990
40	Kr^+	Mg	Viel et al.	1970
40	Ge^+	Ge	Holmén and Högberg	1972
40	Ne^+, Ar^+, Kr^+, Xe^+	Ge	Holmén et al.	1975a
40	Ge^+	Ge	Holmén et al.	1975b
40	Ge^+	Ge	Holmén et al.	1975c
40	Ar^+	Al	Mischler et al.	1984
40	Na^+	Mg,Al,Si	Benazeth et al.	1988b
1.5–40	$H^+, H_2^+, He^+, O^+, Ar^+$	SS	Alonso et al.	1979
2–40	$H^+, H_2^+, H_3^+, D^+, D_2^+, D_3^+,$ $Ne^+, O^+, O_2^+, CO_2^+, C^+,$ N^+, N_2^+	C,Ni,Ta,Pt,Mo	Cawthron et al.	1969
10–40	H^+, H^0, H_2^0	Al,Ni,Ag,Pt,brass,SS	Morita et al.	1966
10–40	Li^+, Sr^+, Cs^+, Ba^+	Al_2O_3, TiO_2	Stein and White	1972
46.5	many	CuBe	Rudat and Morrison	1978
0.5–50	H^0, He^0, N^0	Cu,AgMg	Barnett and Ray	1972a
1.2–50	$H^+, H_2^+, D^+, D_2^+, He^+,$ $B^+, C^+, N^+, N_2^+, O^+, O_2^+,$ $F^+, Ne^+, S^+, Cl^+, Ar^+,$ Kr^+, Xe^+	Al	Alonso et al.	1980

Table 5.2. (continued)

E (keV)	Ions	Targets	Author	Year
2–50	H^+, H_2^+, D^+, D_2^+	Al,Cu,Ag	Baragiola et al.	1978
2–50	$H^+, H_2^+, D^+, D_2^+, He^+$	Li,Al,Cr,Cu,Ag,Au	Baragiola et al.	1979a
5–50	He^+, Ar^+, Xe^+	Cu,Au	Ferron et al.	1981a
0.7–60.2	$H^+, H_2^+, D^+, D_2^+, He^+, N^+,$ $N_2^+, O^+, O_2^+, Ne^+, Ar^+,$ Kr^+, Xe^+	Mo	Ferron et al.	1981b
4–60	Ar^+	Al,Mo	Ferron et al.	1982
20–60	Li^+, Li^0	Be,Al,Ag,Pt,Au,W, CuBe, MgBe	Bethge and Lexa	1966
7.5–70	Ne^+, Ar^+, Kr^+, Xe^+	Cu	Brusilovsky and Molchanov	1971
30–70	Ar^+, Kr^+	Si	Lebedev and Omel'yanovskaya	1974
40–70	Ar^+	Cu,Al,Mo,Ni	Lebedev et al.	1969
80	Ar^+	Be,B,Mg,Au	Veje	1984
80	Ar^+	Be,Mg,Al,Si	Veje	1988
4–80	Ne^+, Ar^+, Xe^+	Si	Benazeth et al.	1989
5–80	$Ar^+, Ar^{2+}, Ar^{3+}, Ne^+, Ne^{2+}$	Mo	Perdrix et al.	1969b
40–80	H^+	Al	de Ferrariis and Baragiola	1986
19–84	cluster	Cu,W	Beuhler and Friedman	1977a
21–84	Ne^+, cluster	Cu	Beuhler and Friedman	1977b
3–90	$He^+, He^{2+}, Ne^{1+-3+},$ $Ar^{1+-5+}, Kr^{1+-7+}, Xe^{1+-8+}$	CuBe	Schram et al.	1966
15–100	Hg^+	W,SS	Hepworth	1970
20–100	Ne^+, Ar^+, Kr^+, Xe^+	Si	Fontbonne et al.	1970
30–110	$H^+, H_2^+, H_3^+, He^+, HeH^+$	Au	Veje	1982
40–115	$Cl^+, Cl^-, I^+, I^-, O^+, O^-$	SS	Hird et al.	1984
15–150/z	H^+, C^{2+-6+}	Cu	Decoste and Ripin	1979
50–170	D^+	Sc, Er, ScD_2, ErD_2	Wurtz and Trapp	1972
200	H^+, H_2^+, H_3^+	Cu	Hasselkamp and Scharmann	1983b
48–212	H^+	Be,C,Al,Ni,Cu,Pt	Allen	1939
50–225	H^+	W	Ewing	1968
40–250	$(H_2O)_x$, cluster	Cu, Al_2O_3	Beuhler	1983

Table 5.3. Experiments with projectile energies exceeding 250 keV; for special notations see Table 5.1

E (MeV)	Ions	Targets	Author	Year
0.1–0.3	H^+, H_2^+, H_3^+	27	Hippler et al.	1988
0.01–0.32	$Se^+, Se_2^+, Te^+, Te_2^+$	Ag	Svensson et al.	1982
0.01–0.35	$H^+, He^+, Ne^+, Al^+, Ar^+, Kr^+, Xe^+$	Al	Svensson and Holmén	1981
0.1–0.4	$(H_2O)_x$, cluster	Cu	Beuhler and Friedman	1980
0.1–0.4	$H^+, He^+, Ne^+, Ar^+, Cu^+, Kr^+, Xe^+$	Cu	Holmén et al.	1981
0.01–0.4	H^+, H_2^+, H_3^+	Al,Cu	Svensson and Holmén	1982
0.03–0.4	H^+	Nb	Musket	1975
0.03–0.4	$H^+, He^+, Ne^+, Ar^+, Kr^+, Xe^+$	Cu	Svensson et al.	1981
0.5	Ne^+	W	Hasselkamp et al.	1980
0.5	H^+, He^+, Ar^+	Al	Hasselkamp and Scharmann	1982
0.025–0.7	$(H)_x$, cluster	SS,Mo,Au	Chanut et al.	1981
0.1–0.8	H^+, He^+, Ne^+, Ar^+	Al	Hasselkamp and Scharmann	1982
0.1–0.8	H^+, He^+, Ne^+, Ar^+	Al	Hasselkamp and Scharmann	1983
0.075–0.9	H^+, H_2^+, H_3^+, He^+	Au	Hasselkamp et al.	1984
0.075–0.9	$H^+, H_2^+, H_3^+, He^+, Ne^+, Ar^+$	Be,C,Mg,Si,Ti, Nb,Au,Cu,Ag,W	Hasselkamp et al.	1987
0.075–0.9	H^+, H_2^+, H_3^+	C,Mg	Hasselkamp et al.	1987
1	H^+	Be,Al,Cu,Au	Kronenburg et al.	1961
1	H^+	Be,Al,Cu,Au,Ta,Au	Gorodetzky et al.	1963
0.08–1	$H^+, H_2^+, H_3^+, C^+, N^+, O^+, Ne^+, CO^+, Kr^+, Ar^+, CO_2^+, Xe^+$	C	Hasselkamp and Scharmann	1983
0.08–1	$H^+, H_2^+, H_3^+, He^+, Ne^+, Ar^+$	Al,Cu,Ag,W,Au	Hasselkamp et al.	1981
1.1/u	He^{2+}, Ar^{12+}	Al	Koyama et al.	1986
1.1/u	$He^{2+}, N^{6+}, Ne^{8+}, Ar^{12+}$	Al	Koyama et al.	1986
1.1/u	He^+, He^{2+}	Au,Ag,Cu,Al,Be	Koyama et al.	1987
1.1/u	He^{2+}, Ar^{12+}	Al,Cu,Au	Koyama et al.	1988
0.84–1.1/u	$He^+, He^{2+}, N^{2+}, N^{6+}, Ne^{2+}, Ne^{8+}, Ar^{4+}, Ar^{12+}, Xe^{9+}, Xe^{27+}$	Al	Koyama et al.	1988
0.8–1.6	H^+, Ar^+	Au, YBa$_2$Cu$_3$O$_7$	Rothard et al.	1988
0.5–2	He^+	SnTe,PbSe	Hasegawa et al.	1988
0.3–2.5	H^+	Al,Si,Cu,Ge	Foti et al.	1974
0.5–2.5	H^+, D^+	Al,V,Fe,SS,Mo,Nb	Thornton and Anno	1977
6.2/u	$He^{2+}, C^{6+}, N^{7+}, O^{8+}$	Al,Ag	Koyama et al.	1982
4.5–8	$He^{2+}, C^{4+}, C^{6+}, N^{4+}, N^{5+}, N^{7+}, O^{5+}, O^{8+}$	Al,Ag	Koyama et al.	1982
4–12	H^+	Al,Cu,Ag,Au	Koyama et al.	1981
2–14/u	H^+, He^{2+}	Al,Ni	Koyama et al.	1976
5–24	H^+	Al$_2$O$_3$, Au	Borovsky et al.	1988
1–40	$H^+, D^+, {}^3He^+, {}^4He^+$	C,Al,Cu,Mo,Ta	Beck and Langkau	1975
9–63	$Li^{2+}, Li^{3+}, C^{2+-6+}$	Al$_2$O$_3$, Au	Borovsky and Barraclough	1989

5.2.1 Positive Ion Impact

A number of publications have appeared since 1968 on the problem of the energy dependence of the total electron yields γ from *clean* metal and semiconductor surfaces. We mention the work of: at low impact energies Perdrix et al. (1969a), Cawthron et al. (1969), Losch (1970), Cawthron (1971), Baragiola et al. (1979a), Alonso et al. (1979, 1980), Ferrón et al. (1981b), Zalm and Beckers (1984, 1985), Lakits et al. (1989a,b); at medium to high impact energies: Ewing (1968), Musket (1975), Svensson and Holmén (1981), Holmén et al. (1981), Hasselkamp et al. (1981), Koyama et al. (1981), Veje (1982), Hasselkamp and Scharmann (1983c), Hippler (1988). Further references may be found in Tables 5.1–5.3 which however mostly deal with contaminated surfaces. Older work was summarized by Krebs (1968).

The general picture of the energy dependence of the total yields γ which had already emerged from older works is confirmed by the more recent experiments: above the threshold for kinetic emission (see below) γ increases with increasing ion energy, then goes through a maximum and decreases thereafter. A fairly consistent data set which covers a large energy interval extending from several keV to more than 10 MeV is available so far only for proton impact on Al, Cu, Ag and Au (see Fig. 5.1). As can be seen in Fig. 5.1 there is still considerable variance between the data sets of Svensson and Holmén (1981) and Holmén et al. (1981) on one hand and Hasselkamp et al. (1981) on the other near the maximum of the yield curves for aluminum and copper. This partial disagreement must be attributed to differences in target cleanness or topography. Good mutual agreement is observed between the data sets of Baragiola et al. (1979a,b), Zalm and Beckers (1984, 1985) and Lakits et al. (1989a) for proton impact on gold at energies ≤ 15 keV (see Fig. 5.5 of Sect. 5.2.2).

The energy position of the maximum of the yields as well as the magnitude of the total electron yields depend on the projectile-target combination (Sect. 5.3). For protons near the maximum of the yield-curve $\gamma \approx 1 - 3$ for metals and semiconductors.

Though it is common practice to present data as a function of the projectile energy it must be kept in mind that the relevant parameter is the projectile velocity (as in most fields of collision physics). This is especially important if the yields obtained from bombardment by different projectile species are to be compared. Instead of a direct comparison as a function of the velocity the yields are sometimes plotted as a function of the energy per atomic mass unit (E/u, i.e. as a function of $v^2/2$), for bombardment by molecular hydrogen ions also the notation energy/proton is a common practice. Figure 5.2 shows an example of $\gamma = f(\text{velocity})$ for a clean aluminum surface bombarded by different light and heavy ions in the energy range from 1.2 to 50 keV. The yields increase proportional to the projectile velocity in this energy regime.

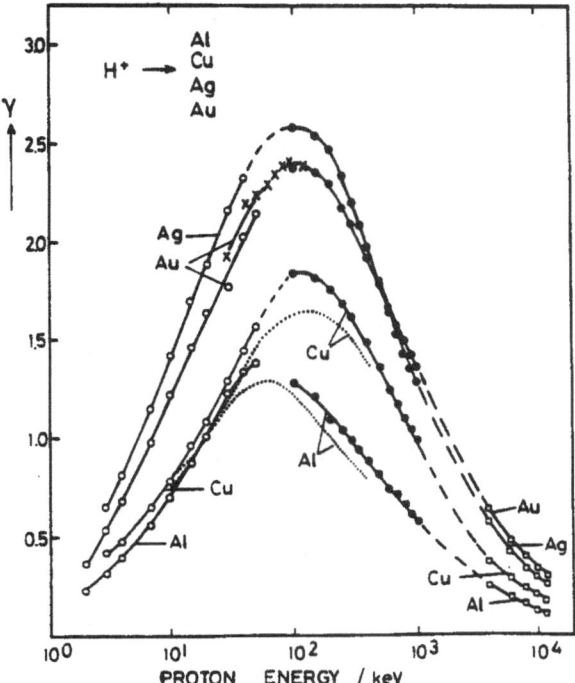

Fig. 5.1. The total electron yield γ as a function of the ion energy for proton impact on clean aluminum, copper, silver and gold: o o o Baragiola et al. (1979b), ● ● ● Hasselkamp et al. (1981), × × × Veje (1982), · · · · Svensson and Holmén (1981) for Al and Holmén et al. (1981) for Cu, □ □ □ Koyama et al. (1981), - - - interpolation between data sets (from Hasselkamp 1988)

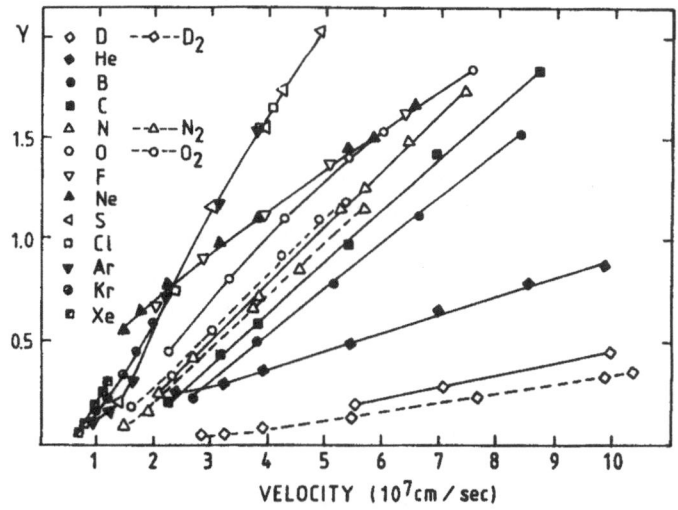

Fig. 5.2. The total electron yields from aluminum under bombardment with different ions as a function of the projectile velocity (from Alonso et al. 1980)

The maximum of the yield-curve was also observed in studies with H_2^+ and H_3^+ and He^+ (Hasselkamp et al. 1981, Hasselkamp and Scharmann 1983c) but so far not for heavy ion bombardment for which the maximum will be shifted to higher impact energies. A decrease of heavy-ion-induced yields at very high energies was observed, e.g., by Linford (1935) and Borovsky and Barraclough (1989).

Older works have reported on an isotope effect which means that different isotopes of the same atomic species should yield different values of γ (e.g. Parilis and Kishinevskii 1960, Arifov 1969, Krebs 1968). It was already noted by Ploch (1951) that the yields from different isotopes have to be compared at the same impact velocity. Nevertheless examples of an isotope effect were reported (e.g. Krebs 1968). If such a phenomenon should exist it should be observed for isotopes with the largest possible mass difference, i.e. for protons and deuterons. No effect was reported in the recent literature (e.g. Large and Whitlock 1962, Dietz and Sheffield 1975, Fehn 1976, Sørensen 1977, Baragiola et al. 1978, 1979a, Ferrón et al. 1981b, Lakits et al. 1989a). Indeed, since electron emission results from an electronic effect, no mass dependence is expected at medium and high projectile velocities. Only near the threshold for kinetic electron emission a small effect may be possible due to differences in projectile retardation and energy straggling (Baragiola et al. 1979a) and to differences in the reflexion coefficient and recoil contribution (Sigmund and Tougaard 1981). Such effects have not been observed so far.

We shortly comment on the fact that kinetic electron emission is a threshold effect. There are many studies on this topic (e.g. Krebs 1968) most of which reported a threshold near a projectile velocity of $0.5–0.6*10^{-5}$ m/s in agreement with the prediction of Parilis and Kishinevskii (1960) (cf. Sect. 4.2). Pre-threshold electron emission was observed for heavy ion impact by, e.g., Ferrón et al. (1981b) and Alonso et al. (1986) (cf. also the article by Varga and Winter in this volume).

5.2.2 Relation of Yields to the Inelastic Stopping Power

The main prediction of classical theories of ion-induced electron emission (Sect. 4.2) is that the total electron yield γ should vary proportional to the inelastic stopping power $(dE/dx)_e$ of the projectile in the target. In order to test this prediction it is convenient to discuss the ratio of the yield and the electronic stopping power (sometimes denoted as "specific yield", Koyama et al. 1976, 1981)

$$\Lambda^{\mathrm{exp}} = \gamma \Big/ \left(\frac{dE}{dx} \right)_e \tag{5.1}$$

(an index is used to distinguish Λ^{exp} from the quantity Λ in the transport theory of Schou 1980 (Sect. 4.2) which, however, has a similar meaning). According to various theories Λ^{exp} should be a material-constant (Sect. 4.2).

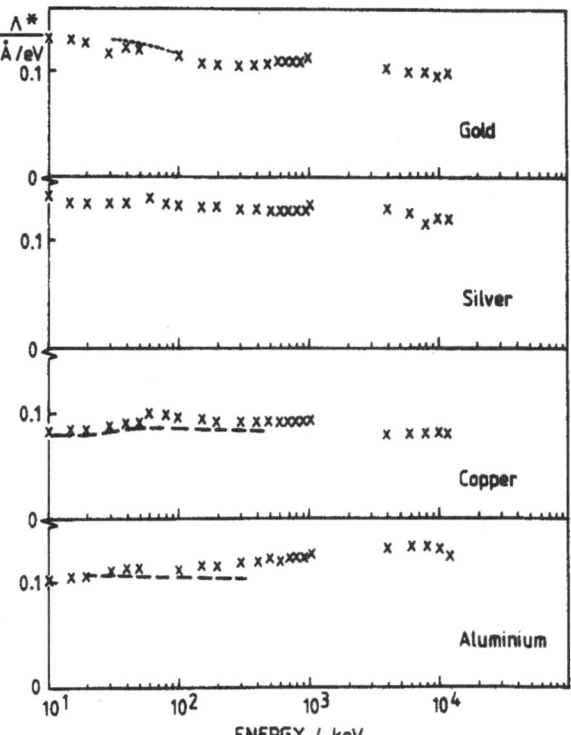

Fig. 5.3. The parameter Λ^{exp} ($= \Lambda^*$) as a function of the proton energy for the yield data of Fig. 5.1; proton stopping powers from Janni (1982) (from Hasselkamp 1988)

Λ^{exp} is easily determined from data of the yields γ and from tabulations of $(\mathrm{d}E/\mathrm{d}x)_e$ (e.g. Northcliffe and Schilling 1970, Anderson and Ziegler 1977, Janni 1982, Ziegler et al. 1985).

In the following proton and heavy ion impact will be treated separately. We consider only the energy dependence (cf. Sect. 5.3 for a discussion of the projectile-target-combination).

Proton Impact. In Fig. 5.3 Λ^{exp} is plotted as a function of the proton energy as derived from the data sets presented in Fig. 5.1. It becomes clear that for a given target material Λ^{exp} is approximately constant in the large energy range from 5 keV to 12 MeV (see also Koyama et al. 1976). Table 5.4 summarizes mean values of Λ^{exp}; the data for different energy regimes agree to within 10 % for a given target material. Data are also available for the backward emission from thin foils bombarded by protons in the energy regime from 300–1200 keV (Rothard et al. 1989 and in this volume). While Λ^{exp} from this work for copper is in fair agreement with the data in Table 5.4, the value for aluminum is much larger possibly due to an insufficient cleanness of the aluminum foil. Still we note that a constant value of Λ^{exp} over a large energy interval emerges also from experiments on thin foils for proton impact (e.g. Rothard et al. 1989, 1990a, Clouvas et al. 1989, see also Rothard et al. in this volume).

Table 5.4. Comparison of published values of the parameter $\Lambda^{\exp} = \gamma/(\mathrm{d}E/\mathrm{d}x)_e$ for proton impact. A: Baragiola et al. (1979a), 2–50 keV; B: Svensson and Holmén (1981) for Al, 10–350 keV and Holmén et al. (1981) for Cu, 10–400 keV; C: Hasselkamp et al. (1981) 80–1000 keV; D: Koyama et al. (1981), 4–12 MeV; E: theory, Schou (1988)

Target	$\Lambda^{\exp}$ in Å/eV				Theory
	A	B	C	D	E
Al	0.11	0.0995	0.11	0.133	0.099
Cu	0.08	0.0771	0.09	0.082	–
Ag	0.125	–	0.13	0.125	–
Au	0.12	–	0.10	0.098	–

It follows from the above that the energy dependence of γ for proton impact is indeed governed by the inelastic stopping power as predicted by various theories and that the quantity $\Lambda^{\exp}$ has the meaning of a material parameter. An approximate value for $\Lambda^{\exp}$ is (Baragiola et al. 1979b)

$$\Lambda^{\exp} \approx 0.1 \ \text{Å/eV} \tag{5.2}$$

an estimate that is correct to within 30 % for a large variety of metal and semiconductor targets as follows from yield data of 27 materials by Hippler et al. (1988).

A calculation of $\Lambda^{\exp}$ is not possible within the framework of semi-empirical theories (Sect. 4.2). A transport-theoretical calculation of the ratio $\gamma/(\mathrm{d}E/\mathrm{d}x)_e = \beta \Lambda$ by Schou (1988) (cf. (4.16) in Sect. 4.2), however, is in good agreement with the experimental data for aluminum at 100 keV impact energy (Table 5.4). Contrary to the theoretical expectations (Sternglass 1957, Schou 1980, Brauer and Rösler 1985, Baklitskii and Parilis 1986, Rösler and Brauer 1989, 1991) there is no experimental evidence for a decrease of $\Lambda^{\exp}$ with increasing energy (Fig. 5.3). This discrepancy can not be explained at present.

The proportionality between γ and $(\mathrm{d}E/\mathrm{d}x)_e$ may be tested in a different way for projectile energies well above the maximum of the stopping power curve. The Bethe formula for the inelastic stopping power may be written as (higher order correction terms are neglected, Sigmund 1975, Chu 1978)

$$\left(\frac{\mathrm{d}E}{\mathrm{d}x}\right)_e = \frac{KZ_2}{E} \log\left(\frac{4m_e E}{m_p I}\right) \tag{5.3}$$

where K is a constant for a given projectile, Z_2 the atomic number of the target material, m_e, m_p the masses of an electron and a proton, respectively, I the mean ionization energy of a target atom and E the projectile energy. With the assumption $\gamma = \Lambda^{\exp}(\mathrm{d}E/\mathrm{d}x)_e$ ((5.1)) the following relation is obtained:

$$\gamma E = K Z_2 \Lambda^{\text{exp}} \log \left(\frac{4 m_e E}{m_p I} \right) \quad . \tag{5.4}$$

Thus a plot of γE as a function of $\log E$ should give a straight line with slope $K Z_2 \Lambda^{\text{exp}}$ which cuts the energy-axis at an energy $m_p I/(4m_e)$. This "Bethe-Fano-Plot" is well-known in atomic collision physics (Inokuti 1971). Not only is a proportionality of γ and $(dE/dx)_e$ easily tested, also Λ^{exp} may be extracted and even mean ionization energies may be accessible. So far the "Bethe-Fano-Plot" was only applied for electron induced yields by Parikh and Shimizu (1976) and tentatively for proton-induced yields by Hasselkamp (1985).

We finally note that the proportionality of γ and $(dE/dx)_e$ may provide the basis for the measurement of the energy dependence of the inelastic stopping power for proton impact by means of the corresponding electron emission. It must be kept in mind, however, that the results of such measurements are strictly valid only for the first surface layers of the target.

Heavy Ion Impact. It is generally acknowledged in the literature that with respect to the energy dependence a simple proportionality between the total electron yield γ and the inelastic stopping power $(dE/dx)_e$ is not valid for heavy ion impact. Most studies report a decrease of the quantity Λ^{exp} with increasing projectile energy (Holmén et al. 1979, Svensson and Holmén 1981, Holmén et al. 1981, Hasselkamp et al. 1981, Hasselkamp and Scharmann 1983c), only at low impact energies both an increase and a decrease with increasing impact energy – depending on the projectile-target combination – was observed (Alonso et al. 1980, Svensson et al. 1982). Thus the interpretation of Λ^{exp} (5.1) as a material parameter is no longer valid for heavy ion impact. Some obvious reasons for this discrepancy with respect to proton bombardment are:

1) Electron emission from recoil ionization (Sect. 4.3.1) may contribute in addition to direct ionization. This contribution is strongly related to the elastic stopping power of the projectile.
2) Projectile ionization may change the distribution function of secondary electrons in the solid as may be concluded from changes in the low-energy electron spectra with impact energy for Ar^+-impact (cf. Sect. 5.6).
3) Inelastic stopping powers taken from tabulations are bulk quantities and may be inapplicable near the surface as noted by Svensson and Holmén (1981) and Holmén et al. (1981).
4) The effective charge of the projectile may change within the escape depth of electrons and the inelastic stopping power may therefore differ from the bulk value. Non-equilibrium effects in electron emission from thin foils have been reported (Rothard et al. 1990a).
5) At low impact energies the concept of a "mean" stopping power breaks down and should be replaced by the concept of an ionization cross sec-

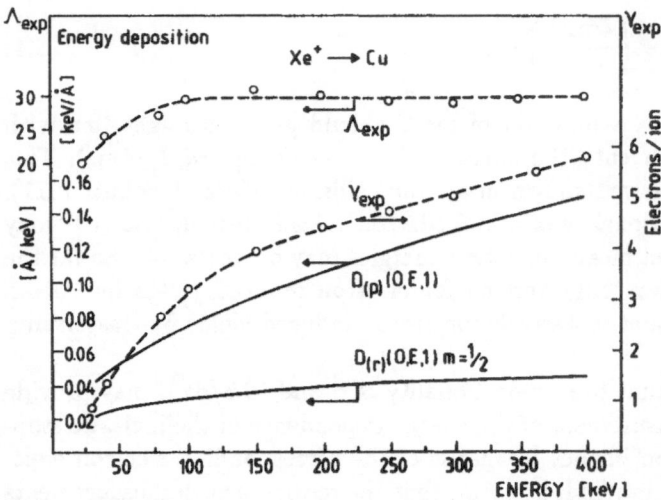

Fig. 5.4. Total electron yields γ_{exp} and the ratio $\Lambda^{exp} = \gamma/D$ including recoil ionization $(D = D_{(p)} + D_{(r)})$ as a function of the energy for bombardment of copper by Xe^+-ions; deposited energy at the surface by direct ionization $D_{(p)}$ and by recoil ionization $D_{(r)}$ (from Holmén et al. 1979)

tion (Schou 1988). Furthermore kinetic electron emission is generally considered to be a threshold effect in contrast to the inelastic energy loss (Alonso et al. 1980). Therefore no simple relation between γ and $(dE/dx)_e$ is expected at low projectile energies.

So far only the problem of recoil ionization has been treated to some extent within the framework of the transport theory developped by Schou (1980). Holmén et al. (1979) have calculated the total energy deposition $D = D_e + D_r$ (4.18) including energy transport by recoiling electrons and found that the ratio γ/D was independent of the impact energy for different ions except for the lowest projectile energies (Fig. 5.4). Svensson and Holmén (1981) and Holmén et al. (1981) performed a simplified calculation which neglects energy transport by recoiling electrons and also reported the ratio γ/D to be independent of the impact energy for different ions. Unfortunately it turned out that the values of γ/D for different projectiles were not the same in contrast to the prediction of Schou (1980) (see (4.18) of Sect. 4.2). No explanation of this discrepancy is available at present.

5.2.3 Impact by Neutral and Negative Particles

Systematic studies of electron emission from clean solid surfaces under bombardment with neutral and negative projectiles are rare and confined to impact energies below 30 keV. Lakits et al. (1989a) performed a comparative study of the total yields γ induced by H^+, H^0 and H^- in the energy regime from 1 to 16 keV (Fig. 5.5). These authors found that $\gamma(H^+) > \gamma(H^0)$ in

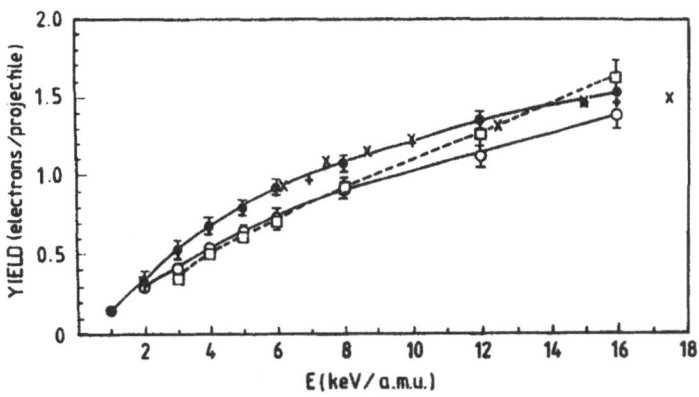

Fig. 5.5. Total electron yields from gold by positive, neutral and negative hydrogen projectiles; ● H$^+$, o H^0, □ H$^-$ (Lakits et al. 1989a); + H$^+$ (Baragiola et al. (1979b); × H$^+$ (Zalm and Beckers 1984); (from Lakits et al. 1989a)

the whole energy range. An explanation was offered in terms of a different inelastic stopping power for both projectiles. It was assumed that the bound electron of H^0 acts by screening the proton charge which leads to a lower effective charge of the H^0-atom compared to the bare proton. A model calculation by Lakits et al. (1990) which takes into account electron loss and capture processes in the first layers of the solid confirmed the expectation of a lower inelastic stopping power of the H^0-atom and consequently a lower total electron yield. The same authors also measured the ratios of the yields of neutral and singly charged noble gas ions (Lakits et al. 1990) and also found a lower yield for neutral atoms than for charged ions in agreement with the concept of screening given above. An early study by Morita et al. (1966) for electron yields induced by H$^+$ and H^0 is in qualitative agreement with the results of Lakits et al. (1989a) and Lakits and Winter (1990). Layton (1973) reported no distinct difference between H$^+$- and H^0-impact below 5 keV.

No results are available for clean surfaces at high impact energies. On the basis of the results for molecular ion impact (Sects. 4.3.4, 5.2.4) one must assume that electron loss processes will contribute to the total yield of a neutral particle at high impact energies (e.g. $\gamma(\mathrm{H}^0) = \gamma(\mathrm{H}^+) + \gamma(e^-)$) both by direct reflexion of the "extra"-electron from the surface (electron loss) and by additional production of secondary electrons. Thus it must be expected that the electron yield induced by neutral particles will be higher than by singly charged ions at high impact energies (above approximately 100 keV for H$^+$ and H^0). A calculation was performed by Gosh and Khare (1962, 1963). The results are in agreement with the expectation. $\gamma(\mathrm{H}^0) > \gamma(\mathrm{H}^+)$ was also observed experimentally by Stier et al. (1954) for bombardment of a "gassy" brass target above 60 keV.

A large variety of measurements have been performed with H$^+$ and H^0-projectiles on contaminated surfaces (e.g. Thomas 1985 and Sect. 6). In contrast to clean surfaces it was found that $\gamma(\mathrm{H}^0) > \gamma(\mathrm{H}^+)$, a result which is not understood at present (Lakits et al. 1989a). A similar result was also obtained by Schackert (1966) for argon- and by Bethge and Lexa (1966) for lithium-projectiles (more of the older work was summarized by Carter and Colligon 1968). We finally note that the neutral yield was found to be smaller than the ion yield for the bombardment of alkali halides by noble gas atoms and ions, but the opposite case was reported by the same authors for bombardment by sodium and potassium projectiles at low impact energies (Arifov et al. 1971a,b).

Electron emission by bombardment of negative ions was only seldom measured. Figure 5.5 contains also the yield vs. energy-curve for H^- on gold by Lakits et al. (1989a). These authors find $\gamma(H^-) \approx \gamma(H^0) < \gamma(H^+)$ at impact energies below ≈ 8 keV (in agreement with older results by Mahadevan et al. (1965) who reported $\gamma(H^-) < \gamma(H^+)$). At energies > 8 keV the yield from H^- is larger than that by neutral hydrogen impact and at 16 keV $\gamma(H^-) > \gamma(H^+) > \gamma(H^0)$. Perdrix et al. (1969a) have reported $\gamma(H^-) > \gamma(H^+)$ for bombardment of molybdenum in the energy range from 5–30 keV. The following explanation was offered by Lakits et al. (1989a) and Lakits and Winter(1990): It is assumed that the outer electron of the H^--ion is readily detached when the ion penetrates the surface, thus $\gamma(H^-) = \gamma(H^0) + \gamma(e^-)$. At impact energies below 8 keV (corresponding to an equivalent energy of the detached electron of ≈ 4 eV) $\gamma(e^-)$ is assumed to be negligible and thus the yields from H^- and H^0 should be the same in the low energy range. At higher energies the equivalent energy of the detached electron increases and causes electron emission by itself which explains the observation of Fig. 5.5. At still higher energies electron loss of both electrons must be expected which should result in a considerable higher yield from H^- than from H^+. No measurements have been performed at high impact energies with H^- on clean surfaces.

Heavy negative ion bombardment was studied by Perdrix at al. (1969a) in the energy range from 5 keV to 30 keV for O^- and Cl^- on Mo (Fig. 5.6) and by Hird et al. (1984) in the energy range from 40 to 115 keV for O^-,

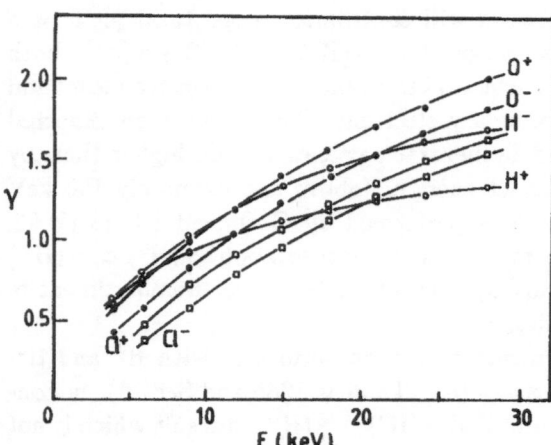

Fig. 5.6. Total electron yields from molybdenum induced by positive and negative ions of H, O, Cl (from Perdrix et al. 1969a)

Cl⁻ and I⁻ on stainless steel. Perdix et al. (1969a) report a lower yield for negative than for positive heavy ion impact in agreement with the hydrogen data discussed above. Hird et al. (1984) did not find any relevant difference between positive and negative ion bombardment. However, it is not clear whether a clean surface was investigated in this experiment.

Gosh and Majumdar (1980) tried to calculate the total electron yield by negative ion bombardment with inclusion of the detachment process. The results are lower than the corresponding data of Perdrix et al. (1969a).

5.2.4 Impact by Molecular Projectiles

Many authors have reported on a molecular effect, which means that the electron yields from a molecular ion differs from the sum of the yields of the constituent atomic *ions* at the same projectile velocities (cf. Sect. 4.3.4). Experimental studies up to 1968 were discussed by Carter and Colligon (1968) and will not be considered here though the general picture is the same as discussed below.

We first consider data for impact by molecular hydrogen ions. At impact energies below ≈ 100 keV/proton $R < 1$ (defined in (4.29)) was reported by e.g. Baragiola et al. (1978, 1979a), Alonso et al. (1980), Chanut et al. (1981), Hasselkamp et al. (1981), Svensson and Holmén (1982), Zalm and Beckers (1984), Lakits et al. (1989a,b). No clear results were obtained by Cawthron et al. (1969), only Veje (1982) reported $R \approx 1$ below 50 keV/proton. At impact energies exceeding ≈ 100 keV/proton $R > 1$ was found in the studies of Hasselkamp et al. (1981), Chanut et al. (1981) and Svensson and Holmén (1982), Hasselkamp and Scharmann (1983c). Thus R is found to increase with impact energy from values < 1 to values > 1 with $R = 1$ at projectile velocities of 2–3 a.u. (Krebs 1983b). The magnitude and the energy dependence of R is a function of the target material (Hasselkamp et al. 1981, Krebs 1983b).

A number of explanations have been put forward to understand the "molecular effect" (Sect. 4.3.4). At low projectile velocities Dorozhkin et al. (1974a) considered differences in the potential emission from atomic and molecular ions. Both Baragiola et al. (1978) and Zalm and Beckers (1984) note that potential emission may contribute but cannot explain the observed effects for molecular hydrogen impact. Baragiola et al. (1978) proposed an interference effect due to the correlation of the constituent particles which may lead to differences in the stopping power (possibly in the sense of Svensson and Holmén 1982, as discussed in Sect. 4.3.4), Zalm and Beckers (1984) considered differences in the reflexion coefficients of atomic and molecular ions.

Presently it seems that simple break-up models as discussed by Svensson and Holmén (1982), Hasselkamp et al. (1987b) and Lakits et al. (1989a,b) suffice to explain the main results. At low impact velocities these authors assume a break-up as $H_n^+ = (n-1)H^0 + H^+$ and a ratio $R < 1$ is just explained by the difference of the stopping powers of H^0 and H^+, respectively, due to the screening effect of the electron of a neutral atom (see also Sect. 4.3.4). This view is strongly supported by the work of Lakits et al. (1989a,b) and Varga and Winter (this volume) who were able to show that the total yield $\gamma\,(H_n^+)$ was equal to the sum of the independently measured yields $(n-1)\gamma\,(H^0) + \gamma\,(H^+)$. These authors demonstrated further that the statistical distribution for impact by H_n^+ was equal to the sum of the distributions of $(n-1)H^0$ plus H^+ but different from the sum of just n protons. At high impact velocities a further break–up of of the hydrogen atoms must be considered due to the electron loss process from the projectile which leads to $\gamma\,(H_n^+) = n\gamma\,(H^+) + (n-1)\gamma\,(e^-)$ (Sect. 4.3.4). The "extra" electron(s) now contributes to the total yield both by the part that is reflected and by exciting secondary electrons by itself. Thus the ratio R may become larger than one at high velocities. Loss electrons were directly observed for impact by molecular hydrogen ions by Hasselkamp et al. (1984) (see Fig. 5.14 of Sect. 5.6). We note that the "break-up" model was already discussed in early studies of electron emission (e.g. Hill et al. 1939, Gosh and Khare 1962, 1963, also Carter and Colligon 1968).

In Sect. 4.3.4 a scaling relation (4.34) was introduced which connects the yield measured for impact by H^+, H_2^+ and H_3^+-ions of the same velocity. The scaling relation was found to be valid for the total yields (Fig. 5.7), for the energy distribution curves (spectra) and also for excitation and ionization cross sections in ion-atom collisions (Hasselkamp et al. 1987b). In view of the preceding discussion this scaling is also an evidence for the simple break-up models.

Molecular effects have also been observed for heavy ion bombardment. Mostly a ratio $R < 1$ is reported for atomic and di-molecular ions of nitrogen and oxygen (Baragiola et al. 1979a, Alonso et al. 1980, Hasselkamp and Scharmann 1983c) and also for CO^+ and CO_2^+ (Hasselkamp et al. 1983c).

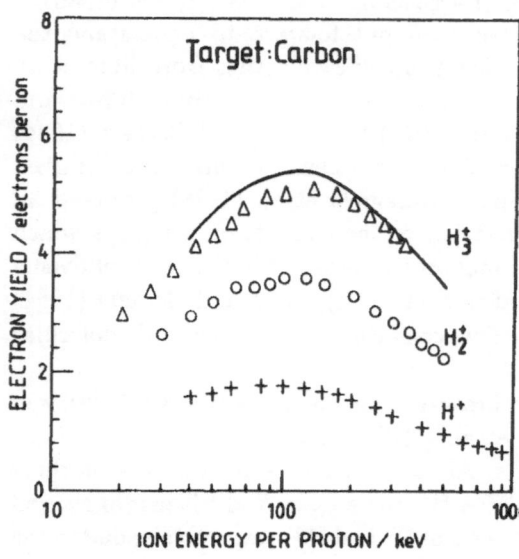

Fig. 5.7. Total electron yields induced by hydrogen ions (data from Hasselkamp and Scharmann 1983c); the yields from bombardment with H_2^+ and H_3^+ are not just two or three times the proton yield at the same projectile velocity; the solid line gives the calculated values for H_3^+ according to the scaling relation (4.34); (from Hasselkamp 1985)

Svensson et al. (1982) studied electron emission by Se_2^+ and Te_2^+ and found $R \approx 1$ at the lowest impact velocities and $R < 1$ at higher velocities. Veje (1981) investigated the bombardment with Sb_n^+ ($n = 1-3$) and observed $R < 1$ for the electron yields which is very different from the respective sputtering coefficients which show a remarkable enhanced emission from bombardment with heavy molecular species.

Strong molecular effects for the backward electron yields – similar to the ones mentioned above – have also been observed in experiments with energetic ions traversing thin solid foils (e.g. Kroneberger et al. 1988, 1989 and Rothard et al. 1990b and this volume).

5.2.5 Cluster Impact

Electron emission by impact of larger molecules, macro-molecules or metal- and water-clusters was studied by several groups (Tables 5.1–5.3). Most authors confirm the validity of the "additivity rule", i.e. the total electron yield is given by the sum of the yields of the constituent particles (or in other words "no enhanced emission" is observed, Thum and Hofer 1979, Hofer 1980). This might be surprising in view of what was said about the molecular effect in Sect. 5.2.4. It must be noted that the velocities of large molecules will in general be small because of their large mass and that molecular effects may be small or absent. Lack of additivity was only reported by Moshammer and Matthäus (1989), by Beuhler (1983) for very large water clusters imping on oxyde surfaces, and by Chanut et al. (1981) for hydrogen cluster ions containing up to nine protons.

It seems to be a general additional result of these studies that pre-threshold emission of electrons is observed (e.g. Staudenmeier et al. 1976, Beuhler and Friedman 1977b, Nikolaev et al. 1978, Beuhler and Friedman 1980, Beuhler 1983). An explanation for this observation may be the possibility that large molecules are in a highly excited (maybe metastable) state before hitting the surface. This energy may be transferred to electrons of the molecule or of the solid which finally leads to electron emission, e.g., by Auger deexcitation. In this connection an experiment must be mentioned in which electron emission was observed by neutral cluster bombardment at an impact energy of only 10 eV (Even et al. 1986). This result cannot be explained by kinetic electron emission.

5.3 Dependence of the Total Yields on the Projectile–Target–Combination

Systematic investigations are now available about the dependence of the total electron yields γ on

1) the type (or atomic number) of atomic projectiles hitting one and the same target which led to the observation of so-called "Z_1-oscillations";

2) the type (or atomic number) of the elemental target material for bombardment with a given projectile which led to the observation of so-called "Z_2-oscillations".

Both cases will be discussed separately below. Before we present the results of this section an introductory remark may be necessary. Since only one and the same target material is used in the measurements of $\gamma = f(Z_1)$ the diffusion and the escape processes within the solid may be considered the same for all projectiles in a first approximation and it is mainly the liberation process of electrons in the direct interaction of projectile and target atoms itself which is tested. This holds as long as the inner distribution function of secondary electrons is independent of the type of the impinging projectile (an assumption which may not be valid, see Sect. 5.6, for the dependence of the energy spectra on the type of projectile) and recoil ionization may be neglected. The situation is more complicated if $\gamma = f(Z_2)$ is measured. In this case electron emission is not only governed by the direct interaction process but also by the material dependence of the diffusion and escape processes (Sect. 4.1).

For some time a monotonic dependence of $\gamma = f(Z_1)$ was assumed in agreement with the prediction of Parilis and Kishinevskii (1960) (4.12,13). Evidence for a non-monotonic behaviour emerged, however, from the work of, e.g., Ploch (1951) and Gaworzewski et al. (1974). A pronounced oscillatory dependence which was approximately independent of the bombarded target material was reported by Fehn (1976), Rogaschewski and Düsterhöft (1976) and Staudenmeier et al. (1976) at impact energies around 20 keV and for contaminated surfaces. These authors noted that the observed Z_1-oscillations coincided with those of the respective electronic stopping powers (see also Alonso et al. 1980). A systematic study by Thum and Hofer (1984) for a clean gold target revealed, however, that there is no strict proportionality of the yields and the electronic stopping power: the Z_1-oscillations were much more pronounced than those known for the respective stopping powers. These authors argue that quasi-molecular processes which are dependent on the shell structures of both projectile and target atoms must play a role and further that excitation of projectiles has to be considered. This view was corroborated in further studies of this group (Ferguson and Hofer 1989) in which it was shown that the observed Z_1-oscillations were different for different clean target materials. Maxima of the oscillatory structure were observed in several cases which could be related to level-matching of atomic levels of projectile and target atoms, respectively, which strongly supports the view of the occurrence of quasi-molecular excitations. Furthermore an anti-phasing of electron yields and of stopping powers was demonstrated for some examples as shown in Fig. 5.8 for a palladium target. No final explanation of the Z_1-oscillations is available at present. So far no experiments have been performed at high impact energies.

In older works no dependence of the total electron yield on the target material was assumed (e.g. Massey and Burhop 1952, Sternglass 1957) which

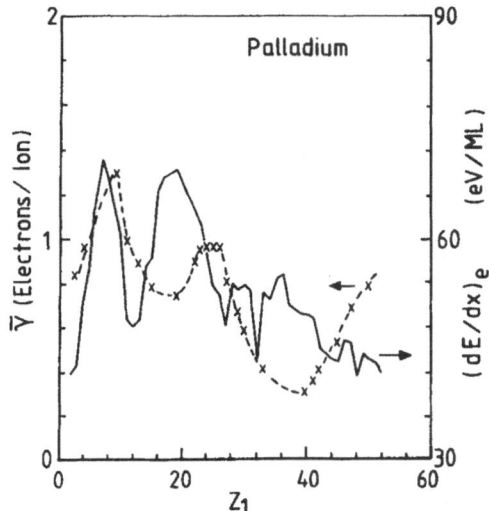

Fig. 5.8. Projectile-dependence of the total electron yield and the electronic stopping power $(dE/dx)_e$, target: Pd (from Ferguson and Hofer 1989)

must be attributed to surface contamination in insufficient vacua. Later a monotonuos increase of γ with the atomic number Z_2 of the target was conjectured (e.g. Large and Whitlock 1962, Bethge and Lexa 1966, Wurtz and Trapp 1972, Dorozhkin and Petrov 1974b, Koyama et al. 1982a, Thomas 1984a). Indications of a non-monotonous behaviour of $\gamma = f(Z_2)$ emerged, however, from the works of Gaworzewski et al (1972/73, 1974), Baragiola et al. (1979a,b) and Hasselkamp (1988). A systematic study is now available which demonstrates a pronounced oscillatory dependence of γ on Z_2 (Hippler et al. 1988, Hippler 1988) at impact energies from 100 keV to 800 keV (Fig. 5.9). These Z_2-oscillations are correlated with the periods of the periodic system and are different from the Z_1-oscillations discussed above. Only small differences were observed for light and heavy ion impact (Hippler 1988). Ex-

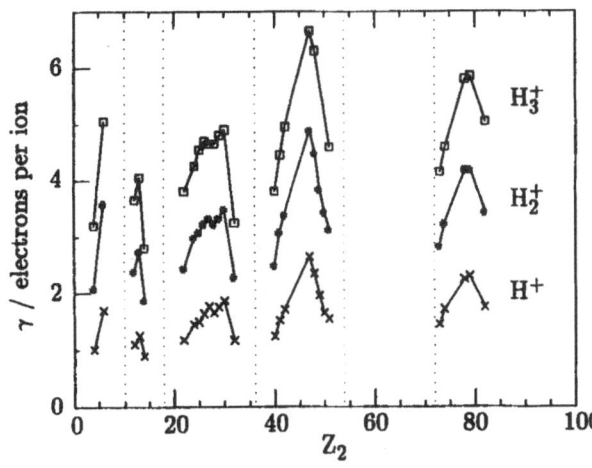

Fig. 5.9. Target-dependence of the total electron yield for hydrogen-ion bombardment (from Hippler et al. 1988)

cellent agreement is found with electron impact experiments as described by Makarov and Petrov (1981). The overall behaviour of $\gamma = f(Z_2)$ is roughly related to the corresponding inelastic stopping powers $(\mathrm{d}E/\mathrm{d}x)_e$ which also show an oscillatory dependence on the atomic number which at high impact energies is mainly due to variations in the number density of the different target materials. However, no strict proportionality between γ and $(\mathrm{d}E/\mathrm{d}x)_e$ is expected since electron yields will also be influenced by the variation of the escape depth and the escape probability for different target materials. It was found that the quantity Λ^{exp} (5.1) also exhibits an oscillatory dependence on the atomic number of the target material. No full explanation of the observed Z_2-oscillations is available at present. Measurements at low impact energies are lacking but would be of considerable interest.

The projectile-target-dependence of electron yields was also investigated for ions traversing thin solid films. A rough proportionality of the total electron yield from the back and the front side of carbon foils with the electronic stopping power for several ions and impact energies was found valid to within $\pm$ 50 % by, e.g., Frischkorn and Groeneveld (1983), Frischkorn et al. (1983) and Rothard et al. (1989, 1990a). Also indications of Z_2-oscillations were observed for protons traversing thin foils of different materials (Clouvas et al. 1989).

5.4 The Total Yield as a Function of the Angle of Incidence

The angle of incidence θ of the ion beam (sometimes also referred to as the tilt angle of the target) is usually measured with respect to the surface normal ($\theta = 0$). If the angle of incidence θ is increased the geometrical pathlength of the impinging particles within the escape zone of secondary electrons is prolonged by a factor $\cos^{-1} \theta$ and the total electron yield is predicted to vary as

$$\gamma(\theta) = \gamma(0) \, \cos^{-1} \theta \tag{5.5}$$

where $\gamma(0)$ is the total electron yield at normal incidence ($\theta = 0$). Equation (5.5) is based on the assumptions of a straight ion path in the solid and of a constant energy deposition over the pathlength within the escape zone.

For not too large θ ($< 70^0$) the dependence predicted by (5.5) was observed experimentally by, e.g., Allen (1939), Aarset et al. (1954) and Gorodetzky et al. (1963) for proton impact, by Evdokimov et al. (1967a), Perdrix et al. (1968) for positive ion bombardment of metal targets and also for negative ion impact by Perdrix et al. (1969a), for semiconductors by e.g. Hasegawa et al. (1988) and for insulating targets by Dietz and Sheffield (1973) and Sørensen (1977). Equation (5.5) is also incorporated in standard theories.

While (5.5) seems to hold strictly for proton bombardment at energies above a few keV deviations were noted by several authors for heavy ion impact (Evdokimov 1967a, Svensson et al. 1981, Svensson and Holmén 1981, Ferrón et al. 1981a, Veje 1984, Benazeth 1987). These authors found that the experimental data for $\theta < 70^0$ could be fitted by an empirical relation:

$$\gamma(\theta) = \gamma(0) \, \cos^{-f} \theta \tag{5.6}$$

where f is a parameter in the range $0.5 \leq f \leq 1.5$ depending on the impact energy, the projectile-target-combination and also on the angle of incidence itself. The parameter f is purely empirical and not related to any theoretical treatment. Several effects have been discussed in order to explain $f \neq 1$ (Ferrón et al. 1981a, Svensson and Holmén 1981, Veje 1984, Hofer 1989):

1) Since the ion pathlength increases with increasing angle θ the slowing-down of the ion must be considered and the assumption of a constant energy loss may break down for large angles of incidence.
2) Recoil ionization may become more effective as the angle of incidence increases because the electron excitation occurs nearer to the surface. This process was studied in some detail by Svensson et al. (1981).
3) The straight-line approximation for the ion path beneath the surface may be inapplicable due to scattering which will increase the effective pathlength.
4) The electron cascade within the solid may be not fully developped for large angles of θ which may lead to a nonisotropic momentum distribution (forward-directed).

Experimental parameters $f < 1$ may then be due to point 1 while $f > 1$ may be explained by points 2 and 3. Unfortunately the experimental results give no clear picture at present, some results are even contradictory: in the energy range from 10–100 keV both $f < 1$ (Evdokimov et al. 1967a, Ferrón et al 1981a, Veje 1984) and $f > 1$ (Svensson and Holmén 1981, Svensson et al. 1981a, Benazeth 1987) have been reported for heavy ion bombardment. Further research is needed both experimentally and theoretically.

For angles of incidence $> 70^0$ the experimental data fall below a curve like (5.6) and eventually even a maximum was observed for heavy ion impact (Fig. 5.10). One reason for the occurrence of a maximum may be that at large angles θ an increasing portion of the ion beam will not penetrate deeply into the solid (and is reflected) and further that planar channeling may occur where the ion beam does not enter the surface at all (Hasegawa et al. 1988). For light ion impact no maximum was observed so far up to angles $\theta = 85°$.

In the work of Dev (1982) an anomalous behaviour was reported for the bombardment of AgMg with He^+ and Ar^+ at low impact energies (≤ 5 keV): a large and broad peak was observed at angles around 75° to 80°. This result is not understood.

We finally note in this section that a non-monotonous dependence of the total electron yield with increasing angle of incidence is observed for mono-crystalline targets. Since crystal effects are not covered by this survey we refer to the review of Brusilovsky (1985).

5.5 The Angular Distribution of Emitted Electrons

There seem to exist only three publications who dealt specifically with the determination of the angular distribution of emitted electrons (differential yield $d\gamma/d\Omega$) at impact energies ≤ 5 keV. From the works of Abbot and Berry (1959), Klein (1965) and Losch (1970) it follows for kinetic electron emission from clean surfaces that the polar angular distribution follows a cos-law like

$$\frac{d\gamma}{d\Omega}(\vartheta) = \frac{d\gamma}{d\Omega}\Big|_{\vartheta=0} * \cos\vartheta \qquad (5.7)$$

where ϑ denotes the angle of emittance with respect to the surface normal. Furthermore no dependence on the azimuthal angle was observed for polycrystalline targets. A cos-shaped polar angular distribution is expected for an isotropic momentum distribution function of electrons inside the solid (Sect. 4.1.2). The emission characteristic is not altered by a planar or spherical potential barrier at the surface (Hofer 1989 and Sect. 4.1.3).

The above results are corroborated by measurements of the doubly differential yield $d\gamma/dEd\Omega$ by Mischler et al. (1984, 1986) for argon impact on aluminum at several keV impact energy. These authors measured the polar angular dependence for background and Auger-electrons. For the background electrons (i.e. the "true" secondary electrons) also a cos-shaped emission was observed. Model calculations have shown that this result is largely independent of the surface topography (Banouni et al. 1985, Mischler et al. 1986, Mischler 1987, Benazeth et al. 1989).

Deviations from (5.7) were only reported in two cases: i) Cawthron et al. (1970) observed a strongly forward directed polar angular distribution for an angle of incidence of 55°; for large angles of incidence of the primary ion beam the assumption of an isotopic momentum distribution inside the solid may fail. ii) For contaminated surfaces structures may be superimposed on the cos-shaped distribution as reported by Losch (1970) and Soszka and Soszka (1980, 1983a); this result is not fully understood.

5.6 Energy Spectra of Emitted Electrons from Clean Metal Surfaces

Energy distribution curves of emitted electrons (or energy spectra) were measured by several authors: either the singly differential yield $d\gamma/dE$ or the doubly differential yield $d^2\gamma/dEd\Omega$ were determined. It must be noted that the results may be different for the two modes of operation: i) it is not clear if the general shape of the electron spectrum is dependent on the angle of emission ϑ (Dorozhkin et al. 1976b); ii) especially for electrostatic analyzers

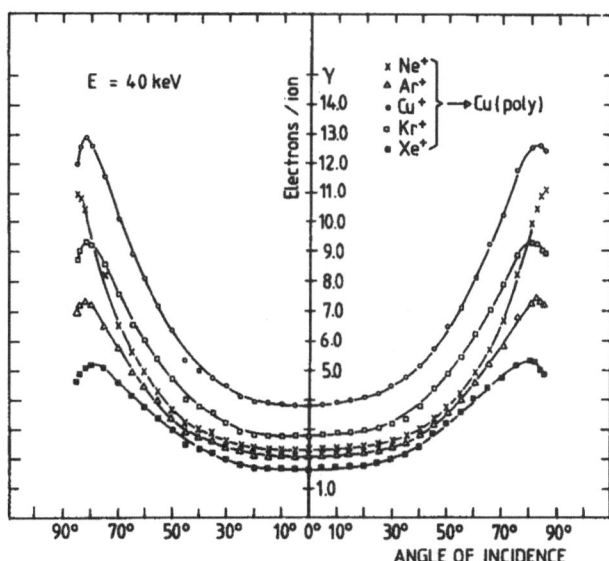

Fig. 5.10. The total electron yield as a function of the angle of incidence for different ions hitting a copper target (from Svensson et al. 1981)

an energy dependent transmission and detection efficiency may influence the measured energy spectrum at low electron energies. In the following section we will first concentrate on the general shape of the electron spectra both at low and high electron energies and will afterwards discuss special features.

Since the early experiments of Füchtbauer (1906) it is well documented that the energy distribution curve increases with electron energy at low energies (in fact starting at zero), exhibits a maximum around 2–4 eV (sometimes called the cascade maximum) and falls off towards higher electron energies (Sect. 4.1). Modern measurements on clean metal surfaces in general confirmed the results of the older investigations but led also to a more refined picture for ion impact energies above 100 keV (Hasselkamp et al. 1987a):

- The shape of the low-energy part of the electron spectrum depends on the target material (Fig. 5.11). The shape may be characterized by the energy position of the low-energy maximum E_{max} and the width of the spectrum at half maximum $\Delta_{1/2}$. For the target materials of Fig. 5.11 E_{max} varies between 1.8 and 3.6 eV and $\Delta_{1/2}$ between 5.4 and 11.8 eV for proton impact (Table 5.5).
- For light ion impact the shape of the low-energy spectrum does not depend on the impact energy.
- For heavy ion impact the shape of the low-energy spectrum does depend on the ion energy: $\Delta_{1/2}$ increases with impact energy (see also Wehner 1966, Lebedev et al. 1969) and is somewhat larger than for proton impact at high ion energies.

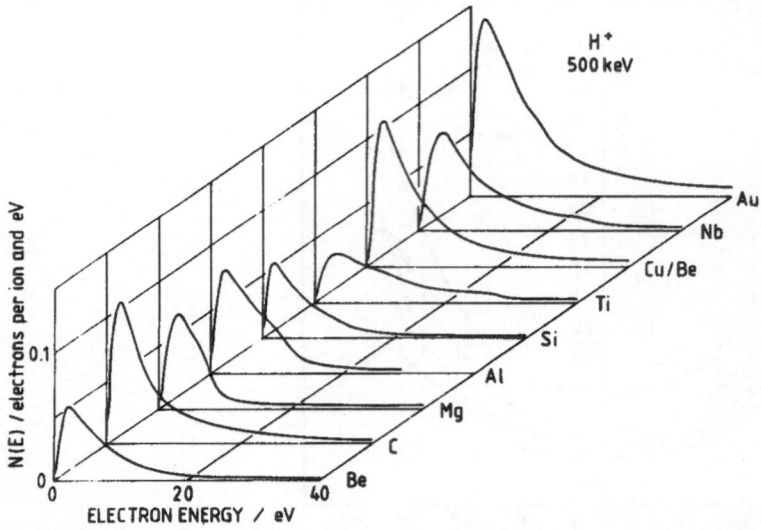

Fig. 5.11. Low-energy spectra $N(E) = d\gamma/dE$ induced by protons at 500 keV from different clean metals (from Hasselkamp et al. 1987a)

Table 5.5. The energy position of the maximum E_{max} and the width of the spectrum at half maximum $\Delta_{1/2}$ for proton-induced spectra of different clean metals (from Hasselkamp et al. 1987a)

Target	E_{max}[eV]	$\Delta_{1/2}$[eV]
Be	2.0	6.8
C	2.0	5.4
Mg	3.0	6.0
Al	2.0	8.2
Si	1.8	6.2
Ti	3.4	11.8
Cu/Be	2.4	6.8
Nb	3.6	9.0
Au	2.2	8.4

These results correct an old opinion which claimed the energy spectrum to be largely independent of the projectile-target combination and the impact energy. Within the theory of Schou (1980, 1988) the shape of the low-energy spectrum is governed by the low-energy stopping power and the surface barrier (Sect. 4.2). Thus a dependence on the target material is expected. However, the shape should be independent on the type and energy of the projectile in agreement with the results for light ion impact but in contrast to the observations for heavy ion bombardment. One possible cause for this discrepancy for heavy ion impact may be that electron excitation from the projectile is not explicitely considered in the theory. The results of Fig. 5.11 and Table

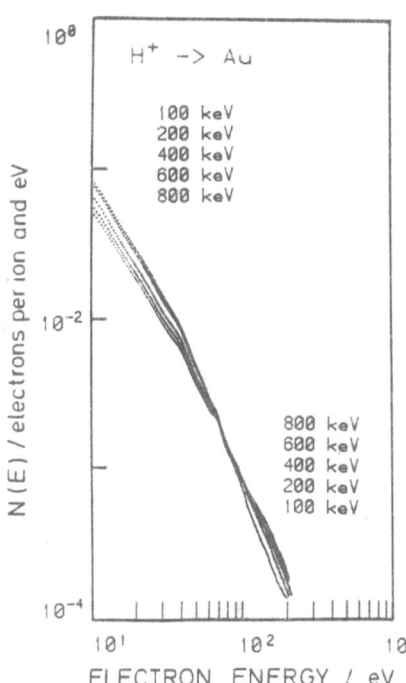

Fig. 5.12. Proton-induced energy spectra beyond the low-energy maximum from gold for different impact energies (from Hasselkamp 1985)

5.5 may be partly compared with spectra obtained under electron bombardment. Bindi et al. (1979, 1980) reported a position of the maximum E_{max} at lower energies and a smaller half width $\Delta_{1/2}$ which furthermore decreased with increasing impact energy.

Beyond the low-energy maximum the energy spectra decrease rapidly (Fig. 5.12). This fall-off can not be described by a simple power law. For high-energy proton impact on gold the shape of the low-energy part (≤ 50 eV) is demonstrated to be independent of the impact energy. At electron energies above 50 eV the slopes of the spectra become impact-energy-dependent (Fig. 5.12): with increasing impact energy the number of energetic electrons is increased. This behaviour may indicate that the emission of low-energy electrons (up to 50 eV for the examples of Fig. 5.12) is indeed the result of a cascade process (Schou 1988) while the high-energy part of the spectrum reflects the direct energy transfer process from the impinging ion to an electron within the solid. The energy spectrum may extend to very large energies: Kronenberg et al. (1961) and Gorodetzky et al. (1963) detected electrons up to an energy of 2 keV for proton impact at 1 MeV. This energy corresponds approximately to the classical high-energy limit for maximum energy transfer (4.35).

The mean energy of emitted electrons

$$< E >= \frac{1}{\gamma} \int E \frac{d\gamma}{dE} dE \tag{5.8}$$

may be estimated from the data in Figs. 5.11,12. For a gold target the mean energy increases from about 15 eV at 100 keV impact energy to about 20 eV at 800 keV impact energy for both proton and argon ion bombardment. Furthermore 50 % of all emitted electrons have energies below 10 eV.

Many authors have reported that a fine structure may be superimposed on the energy distribution curves. While older measurements are of little value due to insufficient surface conditions or possible instrumental artefacts, three different sources of structures are now identified:

- Auger electron emission from target atoms and projectile particles may give rise to humps or even sharp lines in the spectra. This topic is excluded from the present survey and we refer to the references given in Sect. 4.3.5. Auger electron emission from projectiles was studied by e.g. Zampieri et al. (1984), Zampieri and Baragiola (1986) and Benazeth et al. (1988b).
- Shoulders in the low-energy spectrum from nearly free electron metals have been identified as due to one-electron decay of surface and volume plasmons (Sect. 4.3.2).
- Broad structures may result from electron loss from the projectile (Sect. 4.3.3).

The low-energy electron spectrum of aluminum was studied by Hasselkamp and Scharmann (1982a, 1983a), Koyama et al. (1986a) and Benazeth (1987). Shoulders are observed at about 6 eV and 11 eV (Fig. 5.13a,b). The energy positions of these structures are revealed most clearly in the differentiated spectrum, i.e. in $d^2\gamma/dE^2$. An identification is based on two arguments: i) The energy positions are almost independent on both the energy and the type of the projectile and are the same as observed for electron bombardment (e.g. Everhart et al. 1976, Pillon et al. 1976, Bindi et al. 1979); ii) The energy positions correspond to the expected values for a one-electron decay of surface (at 6 eV) and volume (at 11 eV) plasmons in aluminum, respectively (Sect. 4.3.2). This view is further corroborated by the calculations of Rösler and Brauer (1984, 1988, 1991) which were able to reproduce the shoulder due to the decay of volume plasmons. It is a notable result that also heavy ion impact at velocities well below the threshold for direct plasmon excitation (Sect. 4.3.2) lead to the observation of plasmon structures (Fig. 5.13a,b). It was suggested that besides direct plasmon excitation by the projectile also excitation by fast internal secondary electrons may be possible (Hasselkamp and Scharmann 1982a, 1983a). This explanation is in agreement with the theoretical description by Rösler and Brauer (1984, 1988, 1991).

Structures due to the deexcitation of plasmons have also been observed for magnesium at the appropriate energy position and (very weakly) even for silicon (Hippler 1988). A related experiment with fast ions traversing thin foils was performed by Burkhard et al. (1988).

The process of electron loss from the projectile leads to a broad peak in the electron spectrum (Sect. 4.3.3) which at sufficiently high impact energies

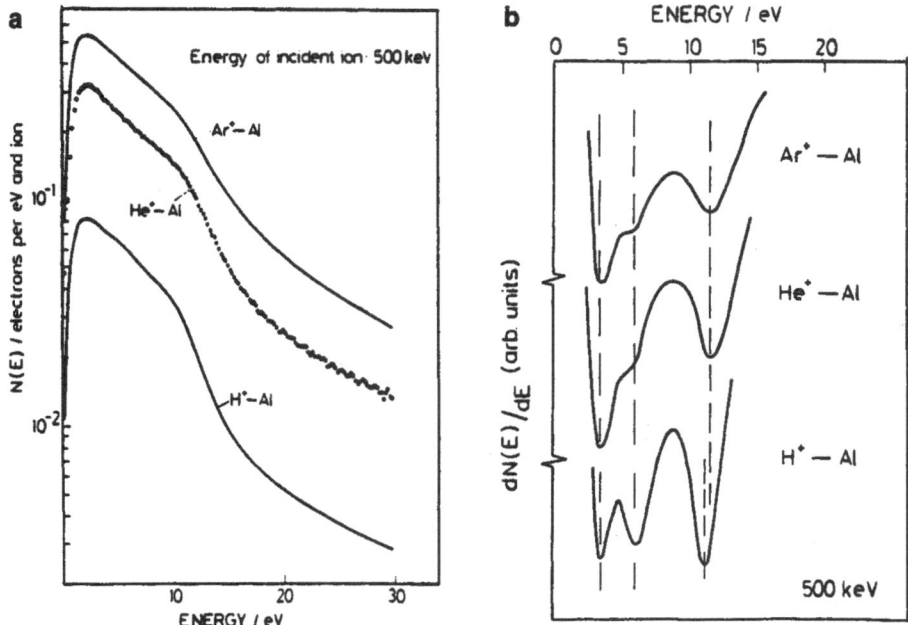

Fig. 5.13. The low-energy spectrum (direct (**a**) and differentiated (**b**)) from aluminum bombarded by different ions at 500 keV ($N(E) = \mathrm{d}\gamma/\mathrm{d}E$); structures are due to one-electron decay of surface and volume plasmons, respectively (from Hasselkamp and Scharmann 1982a)

is approximately centered around the electron energy which corresponds to the velocity of the impinging ion. This feature is most easily detected if the spectrum is compared with the spectrum for impact by a bare nucleus of the same projectile species (Fig. 5.14). For the latter electron loss is an improbable event (an exception will be discussed below). Electron loss has been reported, e.g., for backward emission from thin foils by, e.g., Strong and Lucas (1977), by Oda et al. (1980) and by Suarez et al. (1988) and for emission from solid targets by Hasselkamp et al. (1984) and by Koyama et al. (1987, 1988a). The modification of the electron spectrum by loss electrons results from different processes:

1) Projectile electrons which are reflected at the surface will lead to the loss peak at a projectile velocity (4.27). This peak is easily identified (Fig. 5.14).
2) Electrons lost and reflected below the surface will suffer an energy degradation upon migration towards the surface and will contribute to the spectrum at lower energies.
3) Loss electrons may initiate additional electron emission.

Processes 2 and 3 will always (process 1 only for low impact velocities) lead to a modification of the low-energy part of the spectrum. The contribution

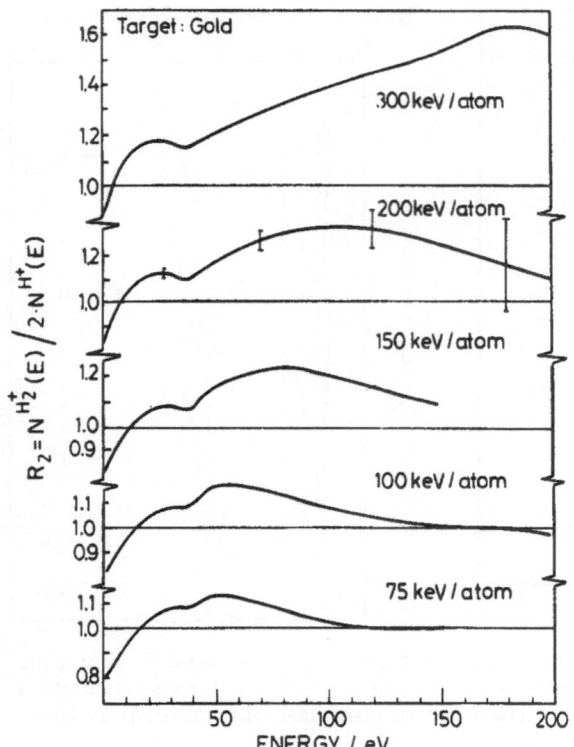

Fig. 5.14. Molecular effects in the ratios (5.9) of the energy spectra induced by H_2^+ and by H^+ from gold at different impact energies; the broad, energy-dependent structure is due to the electron loss process (from Hasselkamp et al. 1984)

of loss electrons to the total electron yield is one of the processes which are responsible for the occurrence of the molecular effect (Sects. 4.3.4 and 5.2.3).

In a first approximation no electron loss process is expected for the impact of bare nuclei. However, a (weak) loss peak was also observed in some cases (Strong and Lucas 1977, de Ferrariis and Baragiola 1986, Suarez et al. 1988). A two-step mechanism was proposed: first the incoming bare nucleus captures an electron into a bound (excited) state; this electron may then undergo the loss process.

The energy position of the loss peak was investigated by Koyama et al. (1987). A weak target dependence was observed. It was explained by the different mean free paths of loss electrons within the solid (see also the article by Rothard et al. in this volume).

Measurements of energy spectra by H^+, H_2^+ and H_3^+ at energies ≥ 50 keV/proton have shed new light on the problem of the molecular effect (Sects. 4.3.4 and 5.2.3). As a first result it was found that the scaling relation (4.34) holds also for the energy spectra including all structures (Hasselkamp et al. 1987b) which may be taken as evidence for the validity of the break-up models discussed in Sect. 4.3.4. New information was gained from the ratios of the energy spectra for H_n^+-ions ($n = 2, 3$) and for protons

$$R_n(E) = \frac{\frac{d\gamma}{dE}(E, \mathrm{H}_n^+)}{n * \frac{d\gamma}{dE}(E, \mathrm{H}^+)} \tag{5.9}$$

as demonstrated in Fig. 5.14. A negative molecular effect ($R(E) < 1$ was observed for low-energy electrons, a positive effect ($R(E) > 1$) at medium-energy electrons and no molecular effect ($R(E) = 1$) for high-energy electrons. An explanation was put forward which is based on the concepts of "close" (small impact parameter) and "distant" (large impact parameter) collisions (Hasselkamp et al. 1984, 1987b). It was assumed that low-energy electrons are mainly excited in distant collisions where screening effects by the accompanying electrons of the molecule ions are active (break-up model (4.33)) which reduces the effective charges of the constituent protons. Energetic electrons were assumed to arise from close collisions where a complete break-up of the molecular ions was anticipated (4.33). In this case the "extra" electrons of the molecular ions contribute to the differential electron yield, e.g. as loss electrons (see above). For energetic reasons projectile electrons can not influence the energy spectrum at energies far above the loss peak which explains the absence of a molecular effect in the high-energy part of the spectra. It must be noted that especially the assumption of a direct emission of low-energy electrons in distant collisions is in conflict with the traditional concept of a pure cascade nature of the low-energy part of the spectrum.

5.7 Non–Metal Targets

Investigations of non-metal targets under controlled experimental conditions are rare. This is especially true for insulating materials for which measurements are difficult due to charging of the target and considerable secondary ion emission.

5.7.1 Semi-Conducting Materials

The energy dependence of the total electron yields and the energy spectra from crystalline Si, Ge, GaAs, GaP for impact energies below 25 keV were studied by Gaworzewski et al. (1972/73, 1974). Yield measurements from crystalline SnTe and PbSe were performed by Hasegawa et al. (1988) with He^+-ions around 1 MeV. Electron yields from amorphous Si- and Ge-targets were measured by Hasselkamp et al. (1987a) and Hippler et al. (1988) in the energy range from 100 to 800 keV. It follows from these experiments that the main features of electron emission which were outlined in the preceding sections for clean metal surfaces are also found for clean semiconductor targets. Neither a notable difference with respect to the absolute values of the yield, nor a different energy dependence of the yields were observed in comparison with the results for metal targets. Also the energy distribution curves resemble those known for metal targets with the exception that the

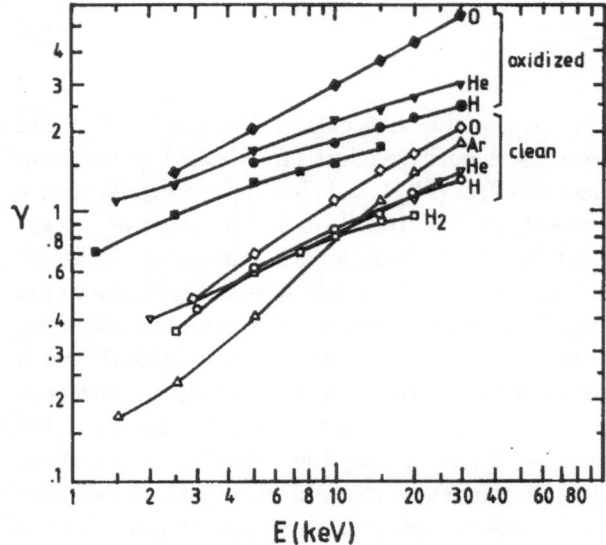

Fig. 5.15. Total electron yields from clean and oxidized stainless steel targets under bombardment with different ions (from Alonso et al. 1979)

position of the low-energy maximum may be shifted somewhat towards lower energies (< 2 eV) (Gaworzewski et al. 1974, see also Table 5.5). No clear correlation between the total yield and the width of the band gap was found by Gaworzewski et al. (1974) for different targets.

5.7.2 Insulating Targets

Only few data sets are available for ion-induced electron emission from clean surfaces of insulating materials (mostly oxidized metals and alkali halides). The results are in qualitative agreement with older measurements (Krebs 1968) and may be summarized as follows (only experiments performed after 1968 are considered):

- The total electron yields are higher than for metal targets (up to a factor 5–10 for the alkali halides; Arifov et al. 1971a,b, Baboux et al. 1971, König et al. 1975, Moshammer and Matthäus 1989). An example for a clean and an oxidized stainless steel surface is given in Fig. 5.15.
- Absolute yields depend on the projectile-target combination. For a given projectile and impact energy the total electron yields seem to be larger for the alkali halides and beryllium oxide than for aluminum oxide.
- The energy dependence of the total yields is similar to that for metal targets. A velocity-proportional increase of γ was observed for impact energies below 40 keV for bombardment of alkali halide targets with noble gas ions (Baboux et al. 1971, Dietz and Sheffield 1975, König et al. 1975), but not for alkali ion impact on aluminum oxide (Dietz and

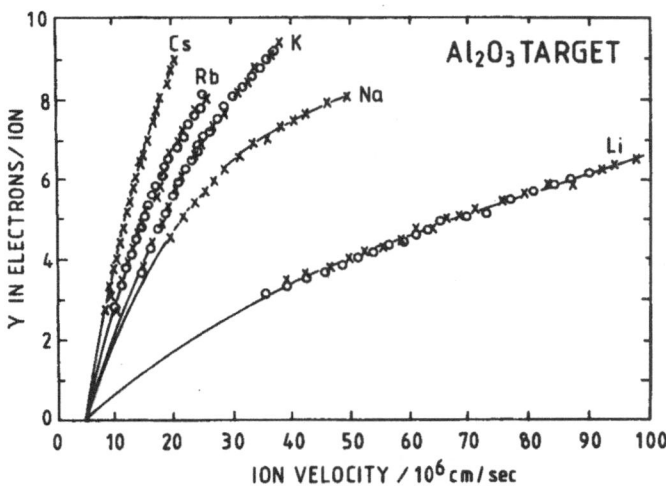

Fig. 5.16. Total electron yields as a function of the projectile velocity for alkali ions bombarding aluminum oxide; × and o denote different isotopes of the same ion (from Dietz and Sheffield 1975)

Sheffield 1975, Fig. 5.16). A decrease of the yields for aluminum oxide was reported at high impact energies (Borovsky et al. 1988, Borovsky and Barraclough 1989). No results are available near the maximum of the yield vs. energy curve.

- For the alkali halides the yields were found to be loosely related to the width of the bandgap E_g:

$$\gamma \propto 1/E_g \tag{5.10}$$

(König et al. 1975).

- Energy spectra were measured by Baboux et al. (1971) and König et al. (1975) for inert gas ion bombardment of alkali halide targets at energies below 30 keV (Fig. 5.17). The shape of the low-energy peak depends on the projectile-target combination and differs from that observed for metal targets: the position of the maximum occurs at lower energies (1-1.5 eV) and the width of the peak is considerable smaller (König et al. 1975).

There seems to be qualitative agreement between the data sets of Arifov et al. (1971a,b), Baboux et al. (1971) and König et al. (1975) for the alkali halides at impact energies below 30 keV. Data from Moshammer and Matthäus (1988), Borovsky et al. (1988) and Borovsky and Barraclough (1989) were taken under unsufficient vacuum conditions. Results from the work of Stein and White (1972) cannot be compared with the data from Dietz and Sheffield (1975) since the former reported the most probable yield from measurements of the electron statistics instead of the average yield γ.

All authors have discussed their results in terms of the semi-empirical theories (Sect. 4.2) from which the difference between metal and insulating targets may be understood from two qualitative arguments:

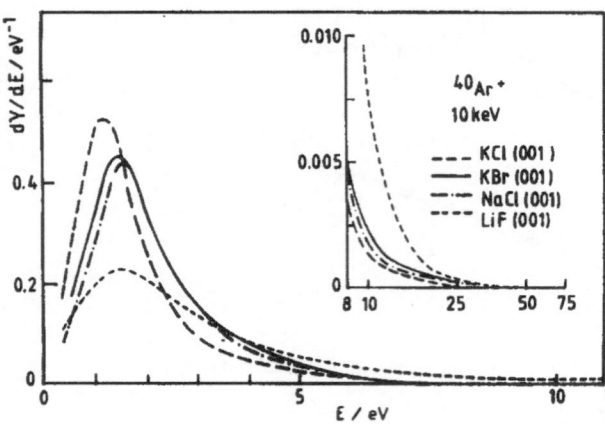

Fig. 5.17. Energy spectra from alkali halide targets under argon bombardment at 10 keV (from König et al. 1975)

1) The mean free path of secondary electrons is larger for insulators than for metals. Hence the escape depth and the yield will be larger (already noted by Bethe in 1941).
2) The work function of insulators is smaller than for metals. More low-energy electrons will be able to surmount the surface barrier. This effect will increase the total electron yield and will shift the position of the maximum towards lower energies.

A quantitative discussion is not possible at present.

A direct comparison between insulating and metal targets is possible for the metal oxides: if the metal target is oxidized only near the surface electron emission may be studied during the process of sputter cleaning thus giving information about the oxide and the clean metal surface one after the other. Several authors have measured both the yields (Fig. 3.1 and e.g. Gaworzewski et al. 1972/73, Alonso et al. 1979, Svensson and Holmén 1981, Ferrón et al. 1981b) and the energy spectra (Hasselkamp and Scharmann 1982b, Hasselkamp 1985, Hasselkamp et al. 1987a, see also Fig. 3.2) as a function of the ion dose. All results confirm the general findings noted above, i.e. a larger yield and a larger amount of low-energy electrons for the oxide state.

Solid gases constitute a special kind of insulating targets. Measurements have been reported for solid hydrogen and deuterium under light ion impact at energies < 10 keV by Sørensen (1976, 1977), Børgesen et al. (1980) and Sørensen et al. (1983). Experimental details may be found in Sørensen (1976). A surprising result of these investigations is the observation of different electron yields from solid hydrogen and deuterium (Fig. 5.18). The authors explained this effect by the differing energy loss of diffusing electrons due to the differing vibrational and rotational energy levels of the molecular solids. The energy dependence of the yields was discussed within the framework of the theory of Schou (Børgesen et al. 1980, Sørensen et al. 1983).

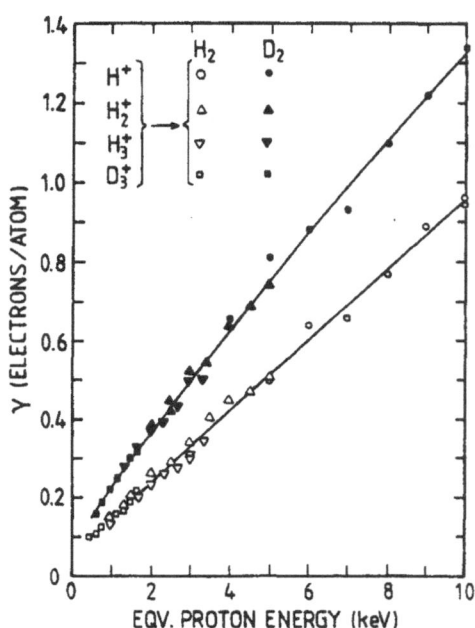

Fig. 5.18. Total electron yields as a function of the energy/proton from solid hydrogen and deuterium bombarded by hydrogen and deuterium ions (from Sørensen 1977)

5.8 Influence of Surface Layers

It is a well-known fact that a contamination of clean metal surfaces leads to a change in the total electron yields (and the energy spectra). This phenomenon was studied in controlled adsorption (or implantation) studies (an early example is the work of Paetow and Walcher 1938). The use of a retarding field analyzer is of great advantage in such studies since besides the measurement of yields and spectra also changes in the work function of the target may be investigated (Sect. 3.3). One may anticipate several different situations:

– Adsorption of up to a monolayer will mainly modify the surface barrier. Both an increase and a decrease of the work function $W = e_0 \Phi$ can be achieved which leads to a decreased or increased escape probability for low-energy electrons (Sect. 4.1.3). Measurements have been performed for monolayer adsorption of H_2 and O_2 on molybdenum (Perdrix et al. 1970), O_2 on tungsten (Hasselkamp et al. 1980) and O_2 on molybdenum (Ferrón et al. 1982). A small decrease of the yields ($\Delta\gamma/\gamma \approx 0.05$) was observed with a simultaneous increase of the work function (Hasselkamp et al. 1980, Ferrón et al. 1982). Adsorption of potassium (Perdrix et al. 1970) or cesium (Coggiola 1986) on a clean molybdenum surface leads to an increased electron emission and a decreased work function (Coggiola 1986). For (sub-)monolayer formation a linear dependence $\Delta\gamma/\gamma \propto -\Delta(W)$ follows from the works of Hasselkamp et al. (1980) (Fig. 5.19), Ferrón et al. (1982) and Coggiola (1986). This simple behaviour indicates that it is indeed the modification of the surface barrier which leads to the observed changes in the yields (and the spectra).

69

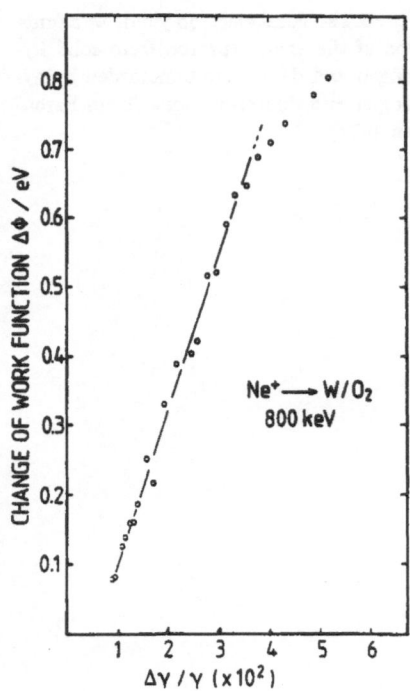

Fig. 5.19. The relative change in the total electron yield $\Delta\gamma/\gamma$ and the change in the surface potential $\Delta\Phi$ of a tungsten target after submonolayer adsorption of oxygen and subsequent bombardment by neon ions at 800 keV (from Hasselkamp et al. 1980)

- Adsorption may lead to a diffusion of foreign atoms into the volume of the surface leading to a modification of the solid at and below the surface. Besides the change in the work function also electron emission from the altered layers must be considered. For the system Al/O_2 an increase of the yield up to 100 % was observed upon adsorption with only a small corresponding decrease of the work function (Ferrón et al. 1982). This result indicates that the emission characteristics are mainly governed by altered sub-surface layers and not by the modified surface barrier. A similar result is expected for adsorption from a steady state background pressure of oxygen which may lead to an oxide formation at and beneath the surface (Veje 1988).
- A modification at and below the surface may also be achieved by implantation of projectiles as follows from the works of Breunig (1972) and Benazeth et al. (1988b).
- Thick deposits of foreign atoms on the surface (film thickness larger than the escape depth) will lead to electron emission which is characteristic of the deposit and not of the underlying metal.

We conclude that the modification of the surface and of near-surface regions is a complicated process which strongly depends on the respective experimental parameters. More systematic work is needed. No general theoretical treatment of this problem is available. Only Schou (1988) has estimated the dependence of the total electron yield on the work function (and the Fermi-energy) for

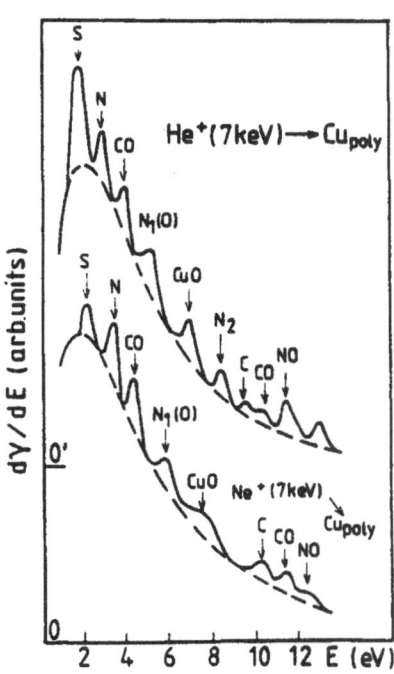

Fig. 5.20. Energy spectrum from contaminated copper under helium and neon ion bombardment; the observed peaks superimposed on the background are labeled with the adsorbed species which may be responsible for their occurrence (from Soszka and Soszka 1980)

nearly-free electron metals. His result confirms the experimental observation of a decreasing (increasing) yield with increasing (decreasing) work function.

In general the energy spectra from targets with defined surface layers exhibit little structure (Hasselkamp et al. 1980, Ferrón et al. 1982) except for features due to Auger-processes (e.g. Benazeth et al. 1988b). This is in contrast to the extensive work of Soszka at ion impact energies < 10 keV on contaminated surfaces in which the doubly differential electron spectra $d^2\gamma/dEd\Omega$ show considerable structure (Fig. 5.20) both with respect to the electron energy and the angle of emission (Soszka 1973a,b, 1978, Soszka et al. 1980, Soszka and Soszka 1977, 1980, 1983a,c, Soszka and Stepien 1981). The authors have attempted to explain the observed features as the result of quasi-molecular ionization processes within the surface layer. The investigation of the fine structure was used as a surface analytical tool by Soszka et al. (1980).

We finally mention an interesting effect observed for negative ion bombardment of contaminated surfaces: the electron yield was shown to increase upon contamination (Perdrix et al. 1969a) indicating that electron detachment may be strongly enhanced by surface layers.

5.9 Impact by Multiply–Charged Ions

We are only concerned with pure kinetic electron emission induced by multiply charged ions. This restricts the discussion to those experiments in which projectile velocities were high enough to prevent potential emission of electrons upon their approach towards the surface (Apell 1987, Varga 1987). We recall that for impact energies below 100 keV no charge effect was observed for kinetic emission by e.g. Schram et al. (1966) and Perdrix et al. (1969b). This is understandable if all ions are neutralized in front of the surface by multiple resonance neutralization. Strong contributions of potential emission have been reported up to medium and high impact energies by e.g. Cano (1973) and Decoste and Ripin (1979). These experiments will not be considered in the following.

If Coulomb interaction prevails the total electron yield should be proportional to the square of the effective charge of the projectile

$$\gamma \propto Z_{\text{eff}}^2 \tag{5.11}$$

for a given target material and the same projectile velocity (cf. Sects. 4.2 and 5.2.1). For a bare nucleus Z_{eff} should equal the electric charge of the projectile as long as charge changing collisions are neglected (electron loss and capture).

Koyama et al. (1976, 1982a) have studied the impact of protons and alpha particles on nickel at energies above 4 MeV/u and indeed found the yields for alpha bombardment to be four times as large as those for protons. For both proton and alpha particle impact a direct proportionality between the electron yield γ and the inelastic stopping power was reported (Koyama et al. 1976). The situation is different for heavier fully stripped ions. In a comparative study of the electron yields induced by He^{2+}-, C^{6+}-, N^{7+}- and O^{8+}-ions of the same velocity (Fig. 5.21) the yields for the highly charged ions were observed

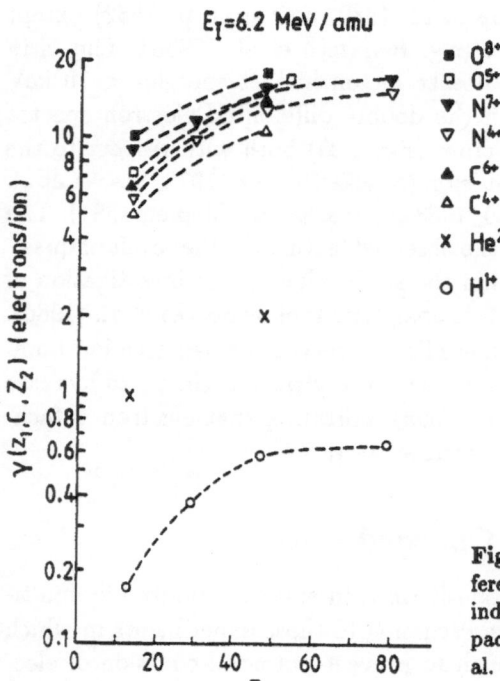

Fig. 5.21. Total electron yields from different target materials (atomic number Z_2) induced by multiply charged ions at an impact energy of 6.2 MeV/u (from Koyama et al. 1982a)

to be lower than expected from (5.11) (Koyama et al. 1982a). It must be noted that (5.11) is only valid if the original charge state is not changed in the solid within the escape depth of electrons. Though the authors expressed their opinion that electron capture was negligible in their experiment such processes may have lowered the effective charge of the ions. Deviations from (5.11) were also observed in a comparison of the yields from partially and fully stripped ions, respectively (Fig. 5.21). In this case the results were partially understood in terms of a modified Bethe formula for the inelastic stopping power of multiply charged ions by Kim and Cheng (1980) which in contrast to the original Bethe theory also considers excitation of the projectiles as a possible energy loss mechanism.

Since experiments on well-defined surfaces are rare we also mention the work of Borovsky and Barraclough (1989) which was performed in high vacuum on oxidized targets. These authors found only a small and irregular dependence of the yields on the charge state of the projectiles which is in disagreement with the results summarized above.

Energy distribution curves have been measured for impact of multiply charged ions at an impact energy of $\approx$ 1MeV/u by Koyama et al.(1986a,b, 1988a,b). The energy spectra $Y(E) = \mathrm{d}\gamma/\mathrm{d}E$ were compared with respect to results obtained for alpha particle bombardment at the same velocity by taking the ratios of the respective spectra (Fig. 5.22). At electron energies above $\approx$ 70 eV the ratios of the differential electron yields correspond approximately to the ratios of the squared electric charges of the respective ions which is attributed to close (unscreened) collisions; from about 20 eV to 60 eV the ratios are reported to be near to the ratios of the respective equilibrium electronic stopping powers; below 20 eV a sharp decrease is observed indicated by a pronounced decrease of the low-energy part of the spectrum for highly charged ions when compared with alpha particles ("depression" of the low-energy spectrum). The last feature was explained by a screening

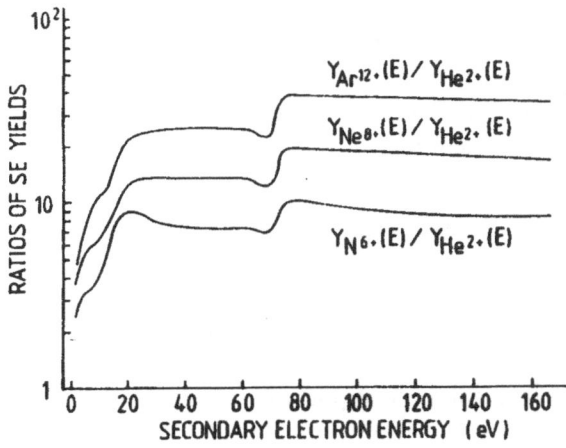

Fig. 5.22. The ratios of the electron spectra induced by N^{6+}, Ne^{8+}, Ar^{12+} and by alpha particle bombardment as a function of the energy of emitted electrons (data points not shown, from Koyama et al. 1986b)

effect by the target electrons in "distant" collisions (Koyama et al. 1986b). The experimental observations and the argumentation are similar to the case of the molecular effect (Sects. 4.3.4, 5.2.4). A notable structure is always observed at the position of an Auger shoulder (near 70 eV in Fig. 5.22). This special feature which is also observed for impact by molecular hydrogen ions (Fig. 5.14) was discussed by Koyama et al. (1988b).

5.10 Temperature Dependence

There is general agreement that the total electron yield does not change with the temperature of the target for *clean* metals (e.g. Simon et al. 1962, Krebs 1968, Perdrix et al. 1968). This is understandable since the mean free path of excited electrons (the only temperature-dependent quantity in the semi-empirical theory, Sect. 4.2) is governed by inelastic electron-electron collisions. In contrast, for insulators a gradual decrease of the yield γ with increasing temperature is expected due to increasing electron-phonon interaction (Baboux et al. 1971 for a NaCl-target).

A special situation is encountered in those cases where the solid undergoes a phase transition at a certain temperature. A change of the electron yield was observed in several cases (Soszka and Stepien 1981, Soszka and Soszka 1983b, Benazeth et al. 1989). Changes of the electron yield were also reported during the annealing of defects in semiconductor crystals (Holmén and Högberg 1972, Lebedev and Omelyanovskaya 1974, Holmén et al. 1975a–c).

In all investigations noted above the temperature of the target was increased starting at room temperature. There is only one notable experiment in which the temperature was decreased below room temperature and which yielded an unexpected result (Rothard et al. 1988). For both a gold foil (electron emission measured in both forward and backward directions) and a thick high-temperature superconductor (backward yield only) the electron yield was observed to decrease with decreasing target temperature (especially for proton bombardment) in disagreement with the above results and with the theoretical expectation (Fig. 5.23). The case of the high-T_c target is especially complicated in that the yield was found to increase again below the critical temperature for superconduction. These results are not understood at present. The possibility of phase transitions in near-surface regions must be considered for the high-T_c superconductor.

5.11 Emission of Polarized Electrons

The spin polarization of *electron*-excited secondary electrons from ferromagnetic surfaces has been intensively studied (Kirschner 1985, Feder 1985). Only one experiment in this field has been performed with low-energy *ion* bombardment (Kirschner et al. 1987). The results are important enough to be shortly summarized though the bombardment energy was below 5 keV and a single

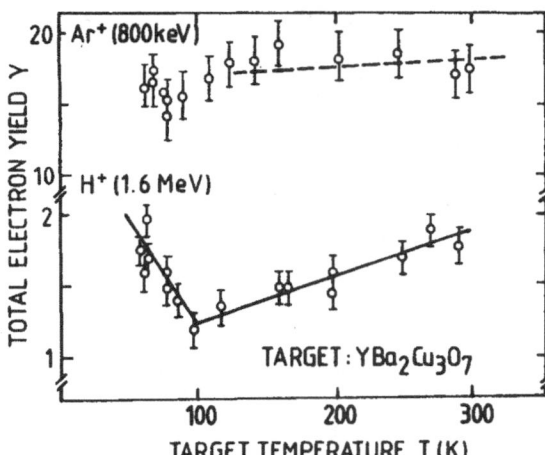

Fig. 5.23. The temperature dependence of the total electron yield from a gold foil and a massive high-temperature superconductor ($YBa_2Cu_3O_7$) under bombardment with high energy protons and argon ions (from Rothard et al. 1988)

crystal Fe(110)-target was used (such experiments were deliberately excluded from this survey in the other sections). Substantial spin polarization was observed (up to 30 %) for impact by He^+, Ar^+ and Xe^+ at all energies from 0.5 to 4 keV. The degree of polarization increased with increasing energy. Spin polarization was observed in both the regions of potential and kinetic emission. An important result is the fact that heavy ion bombardment did not destroy the magnetic structure of the surface. It was shown that sputter profiling of magnetic surfaces is possible, a technique that eventually may allow three-dimensional imaging of magnetic structures (Kirschner et al. 1987).

6. Applied Aspects of Ion–Induced Electron Emission

In this section we will give a short overview of practical applications of ion-induced electron emission. One may distiguish between two aspects:

- In many cases electron emission is an unwanted by-product of energetic particle-solid interactions. Knowledge of the principles of ion-induced electron emission is necessary either to supress the emission (e.g. by suitably biasing of slits or targets) or to consider possible disturbing effects of electron emission (like the charge-up of insulating surfaces or the modification of the plasma sheath in plasma-surface interactions).
- Electron emission may be directly used as an experimental tool, e.g., in the design of particle detectors or in ion microscopies.

In Table 6.1 we give a short summary of studies which are concerned with practical aspects and applications of ion-induced electron emission.

Table 6.1. Summary of experiments dealing with applied aspects of ion-induced kinetic electron emission

Topic	Autor	Year
ion beam current measurement	Matteson and Nicolet	1979
very small ion currents	Blauth et al.	1972
	Krebs	1983a
ion detectors in mass spectrometry	van Gorkum and Glick	1970
	Lao et al.	1972/73
	Pottie et al.	1973
	Rudat and Morrison	1978
detection of clusters	Beuhler and Friedman	1977b
	Nikolaev et al.	1978
	Thum and Hofer	1979
	Beuhler	1983
	Moshammer and Matthäus	1989
neutral particle detectors	Morita et al.	1966
	Fitzwilson and Thomas	1971
	Barnett and Ray	1972a
	Pradel et al.	1974
	Kislyakov et al.	1976
	Ray et al.	1979
	Rinn et al.	1982
ion microscopy	Möllenstedt and Lenz	1963
	Kneis et al.	1982
	Levi-Setti	1983
	Levi-Setti et al.	1985
	Lai et al.	1986
	Levi-Setti et al.	1988
	Allen and Brown	1989
focused ion beam technology	Prewett	1984
	Heimann and Blakeslee	1986
	Melngailis	1987
	Morita and Hashimoto	1988
	Lewis and Glocker	1989
nuclear track formation	Oda and Lyman	1967
	Monnin	1975
	Groeneveld	1988
sputter deposition of films	Chapman et al.	1974
	Hanak and Pellicane	1976
plasma-surface interaction	Hofer	1987
problems in fusion	Barnett and Ray	1972b
	Vernickel et al.	1978
	McCracken and Stout	1979
	Thomas	1984a
	Akazaki et al.	1984
	Thomas	1985

Table 6.1. (continued)

Topic	Autor	Year
ionization cross sections	Benazeth et al.	1976
	Benazeth et al.	1977
	Benazeth et al.	1978
	Hasselkamp	1985
determination of attenuation lengths	Needham et al.	1977
	Powell and Stein	1977
surface analysis	Soszka	1978
	Soszka and Soszka	1980
	Soszka et al.	1980
monitor for a clean surface	Hasselkamp and Scharmann	1982b
	Baragiola et al.	1979a,b
	Svensson and Holmén	1981
	Rothard et al.	1990b
monitor for radiation defects	Evdokimov et al.	1967c
	Holmén and Högberg	1972
	Lebedev and Omel'yanovskaya	1974
	Holmén et al.	1975a–c
monitor for sub-monolayer adsorption	Hasselkamp et al.	1980
alignment of single crystals	Feijen et al.	1973
background in PIXE	Folkmann et al.	1974

Early research with respect to applications were in the field of gas-discharges as reviewed by Massey and Burhop (1952) and Little (1956). Also the effect of electron emission on the problem of high-voltage break-down in accelerator tubes was studied (e.g. Hill et al. 1939, Aarset et al. 1954, Bourne et al. 1955). Another problem (already noted by Füchtbauer 1906) is the exact determination of an ion beam current. Usually an ion current is measured by a suitably designed Faraday-cup where provision must be made that electrons which are emitted as the result of ion bombardment can not leave the cup. We refer to the experimental study of Matteson and Nicolet (1979) in which the design of a Faraday-cup for energetic ions was investigated on the basis of the principles of ion-induced electron emission.

With the development of electron multipliers in the beginning fifties the question of particle detectors became a subject of considerable interest. The gain of open multipliers was studied as a function of the ion energy and ion type in numerous experimental studies (as an early example we cite the work of Barnett et al. 1954). Extensive work was devoted to ion detectors (open multipliers or channeltrons) with respect to mass spectrometry and secondary ion mass spectrometry (SIMS) (e.g. van Gorkum and Glick 1970, Lao et al. 1972/73, Pottie et al. 1973, Rudat and Morrison 1978). Several studies dealt with the identification of cluster ions by means of the total electron yields (Beuhler and Friedman 1977b, Nikolaev et al. 1978, Thum and Hofer 1979, Beuhler 1983, Moshammer and Matthäus 1989). Detection and identification of very large water clusters with mass to charge ratio ≥ 10000 were reported by Beuhler and Friedman (1977b) and Beuhler (1983). Older work with hydrocarbon ions was reviewed by Krebs (1968).

Electron emission provides the means for the construction of simple detectors for beams of energetic neutral particles. In most cases open multipliers, channeltrons or channelplates are used. A major problem is the calibration of the instrument which must be based on an alternative method for the detection of neutral beams (e.g. Hasted 1964, Krebs 1968). For an easy operation a calibration curve based on the ratio of the electron yields of neutral and singly charged positive ions of the same species is of importance and has been measured by several groups (see Table 6.1).

Ion detectors based on the principle of the ion-electron converters (see Sect. 3.2.2) are in use for the determination of very low ion currents (for a review see Krebs 1983a). Currents as low as some 10^{-22} A can be measured.

Ion-induced electrons may be used to form an image of the emitting surface by electron-optical means. Ion electron emission microscopy was reviewed by Möllenstedt and Lenz (1963). A lateral resolution of 20–70 nm was achieved in these instruments. Recent developments in ion microscopy concentrated on the combination of an electron-imaging system in combination with scanning PIXE (proton induced X-ray emission) or scanning SIMS (secondary ion mass spectrometry). A proton microprobe which operated at a proton energy of 2–4 MeV and a minimum spot size of 1–2 μm was developped by e.g. Kneis et al. (1982). Scanning SIMS facilites were described by Levi-Setti (1983), Levi-Setti et al. (1985, 1988) and Allen and Brown (1989). The use of liquid metal ion sources provided a minimum spot size of about 20 nm with gallium ions at about 50 keV. A review of the relationship between electron and ion induced secondary electron imaging was given by Lai et al. (1986). A major advantage of ion electron imaging over the conventional scanning electron microscope is the higher topography contrast. This feature is especially welcome in focused ion beam (FIB) technologies as in ion implantation and mask repair where electron imaging systems are used for a correct positioning of the target (Prewett 1984, Melngailis 1987).

Ion-induced Auger electron spectroscopy (IAES) has been the subject of considerable interest. In principle it may be used as a surface analytical tool like AES (Auger electron spectroscopy with primary electrons). As already mentioned in Sect. 4.3.5 proton and electron induced Auger spectra are equivalent and there is no advantage of using energetic proton bombardment instead of conventional AES. It should be noted, however, that the analysis of Auger features in the electron spectra may provide information of the cleanliness or the state of the surface under proton bombardment without the use of a separate Auger spectrometer. Heavy ion-induced Auger spectra on the other hand often show additional "atomic-like" peaks which were identified to originate from the relaxation of inner-shell excited, sputtered atoms in front of the target surface in vacuum. These peaks are therefore not directly related to the surface and bulk composition of the target.

The measurement of Auger electron yields is a standard procedure for the determination of inner-shell ionization cross sections in ion-atom collision physics (e.g. Stolterfoht 1978). There have been attempts to derive ionization cross sections also from Auger electron yields from ion-bombarded solids e.g. for the K-shell of beryllium (Benazeth et al. 1976), for the $L_{2,3}$-shells of magnesium and aluminum (Benazeth et al. 1977, 1978) and of the $4f$- and $5p$-shells of gold (Hasselkamp 1985). These approaches are hampered by a number of difficulties and the results must be more considered an estimation than a correct derivation of cross sections: i) the structures due to Auger electrons are superimposed on the background of "true" secondary electrons; a background correction is usually applied in the vicinity of the Auger structure; it was demonstrated, however, by Toburen et al. (1982) that Auger structures possess an extending low-energy tail which cannot be neglected; ii) in some of the above mentioned work atomic-like Auger peaks from heavy ion-bombardment were falsely included in the calculation of the Auger yield; however, these peaks are not the result of the primary collision process; iii) the diffusion and escape of Auger electrons is treated within the framework of the semi-empirical theory (Sect. 4.2); the necessary input quantities (i.e. the escape length and the surface barrier) are not well known and must be estimated. It follows from the above that any claims of a good agreement of inner-shell cross sections derived from Auger yields in ion-solid interactions must be considered fortuitous at present. A comparison between experiment and theory on a qualitative basis was recently reported by Benazeth et al. (1988a).

The foregoing problem was turned around by Needham et al. (1977) and by Powell and Stein (1977): these authors used measured Auger yields from proton-bombarded beryllium in order to derive the escape length (or attenuation length) for Auger electrons in the solid. In this case the ionization cross sections were used as the input quantities which were taken from theoretical studies in ion-atom collisions. The evaluation procedure was also based on the semi-empirical theory. This approach is hampered by the same draw-backs as the determination of cross sections.

There is one notable study by Soszka et al. (1980) in which ion-induced electron emission was used as a surface analytical tool. The method called "angle resolved ion-electron spectroscopy" (ARIES) is based on the observation that at bombardment energies $\leq$ 10 keV sharp peaks in the low-energy electron spectra are recorded if adsorbed atoms are involved in the electron creation process (see Fig. 5.20). As the underlying physical process the authors tentatively assume the inner-shell ionization of an adsorbed atom by ion impact and a subsequent Auger-effect involving electrons from the conduction band of the substrate. Adsorbed atoms of carbon, oxygen and sulfur were identified on a copper surface. From the anisotropy of emitted electrons statements about the position of sulfur atoms in the surface were made possible.

The measurement of the total electron yield may give information about changes in the composition, in the structure or in the topography of the surface. Thus ion-induced electron emission was used as a monitor of the cleaning process of a contaminated or oxydized surface (Sect. 3.1.2), of the annealing of radiation effects due to ion bombardment, of phase transitions and of submonolayer adsorption of oxygen on a tungsten surface (see Table 6.1). These applications are purely qualitative so far and are possibly valid only in special cases.

The contribution of secondary electrons on the background in proton induced X-ray emission (PIXE) studies was discussed by Folkmann et al. (1974).

We add one example for the bombardment of single crystals: as shown by Feijen et al. (1973) electron emission may be used efficiently for the alignment of crystals by reference to the anisotropic angular distribution of emitted electrons.

The important role of ion-induced electron emission in such fields as plasma-surface interactions (Hofer 1987), in fusion research (see Table 6.1, yield data relevant to fusion were summarized by Thomas 1984a, 1985), in track formation in solids (Monnin 1975, Groeneveld 1988) and in biological materials (Oda and Lyman 1967, Inokuti 1983) is especially emphasized.

It must be noted that practical considerations prevail in most of the mentioned studies. In general the derived data are for contaminated and "gassy" surfaces and are therefore often not of basic interest.

7. Summary and Outlook

The large number of experiments (and the corresponding increase of available data) which have been described in the present survey documents the progress that has been achieved in the field of ion-induced kinetic electron emission since 1968. It can not be overlooked, however, that systematic investigations on defined surfaces are still rare and that many experiments have simply increased the amount of available data but not the basic understanding of the phenomenons. The following problems should be investigated in more detail:

- Topographic effects and the modification of surfaces under ion bombardment must be studied with respect to their influence on electron emission.
- The basic relationship between the total electron yield and the inelastic stopping power of the projectile in the target is so far only validated for light ion impact but is unclear for heavy ion bombardment. Nonequilibrium effects near the surface are of special interest.
- The projectile-target dependence of the total electron yields and of the electron spectra should be studied in different energy ranges. In the low-

energy range the possible excitation of electrons by quasi-molecular collisions needs further consideration (also with respect to the threshold behaviour of kinetic electron emission).
- The present knowledge about electron emission from insulators (and from compound materials) is unsufficient.

Further points of experimental interest include the molecular effect and the electron loss process for neutral and cluster bombardment, the influence of defined surface layers, and kinetic emission induced by multiply-charged ions at high energies.

Acknowledgements. The continuous interest and support of Prof. Dr. Dr. h.c. mult. A. Scharmann during the preparation of the present survey is gratefully acknowledged. The author wishes to thank Prof. Dr. K.-H. Schartner, Prof. Dr. W. Brauer and Dr. W. Krüger for many helpful suggestions and D. Evers for the critical reading of the manuscript.

References

Aarset, B., R.W. Cloud and J.G. Trump, 1954: J. Appl. Phys. **25**, 1365. Sects. 6, 5.4

Abbot, R.C. and H.W. Berry, 1959: J. Appl. Phys. **30**, 871. Sects. 3.4, 5.5

Abroyan, I.A., M.A. Eremeev and N.N. Petrov, 1967: Sov. Phys. USPEKHI **10**, 332. Sect. 1, (Tab. 1.1)

Akazaki, M., M. Masuda, H. Yano and T. Taniguchi, 1984: Jpn. J. Appl. Phys. **23**, 1155. Sect. 6, (Tab. 6.1)

Allen, J.S., 1939: Phys. Rev. **55**, 336. Sects. 5.4, 6, (Tab. 5.2)

Allen, G.C. and I.T.Brown, 1989: J. Physique Coll. C2 **50**, 121. Sect. 6, (Tab. 6.1)

Alonso, E.V., R.A. Baragiola, J. Ferrón and A. Oliva-Florio, 1979: Rad. Effects **45**, 119. Sects. 3.1.3, 4.1.3, 5.1, 5.2.1, 5.7.2, (Tab. 5.2)

Alonso, E.V., R.A. Baragiola, J. Ferrón, M.M. Jakas and A. Oliva-Florio, 1980: Phys. Rev. B **22**, 80. Sects. 4.1.1, 4.3.1, 5.1, 5.2.1, 5.2.2, 5.2.4, 5.3, (Tab. 5.2)

Alonso, E.V., M.A. Alurralde and R.A. Baragiola, 1986: Surface Sci. **166**, L155. Sects. 5.1, 5.2.1, (Tab. 5.1)

Anderson, H.H. and J.F. Ziegler, 1977: *Hydrogen Stopping Powers and Ranges in All Elements*, (Pergamon Press, New York). Sect. 5.2.2

Apell, P., 1987: Nucl. Instrum. Meth. Phys. Res. B. **23**, 242. Sect. 5.9

Arifov, U.A., 1969: *Interaction of Atomic Particles with a Solid Surface*, (Consultants Bureau, New York). Sects. 1, 3.2.1, 4.2, 5.2.1, (Tab. 5.1)

Arifov, U.A., N.N. Flyants and R.R. Rakhimov, 1971a: Bull. Acad. Sci. USSR, Phys. Ser. **35**, 228. Sects. 5.2.3, 5.7.2

Arifov, U.A., R.R. Rakhimov and S. Gaipov, 1971b: Bull. Acad. Sci. USSR, Phys. Ser. **35**, 516. Sects. 5.1, 5.2.3, 5.7.2, (Tab. 5.1)

Baboux, J.C., M. Perdrix, R. Goutte and C. Guillaud, 1971: J. Phys. D: Appl. Phys. 4, 1617. Sects. 5.1, 5.7.2, 5.10, (Tab. 5.2)

Baklitskii, B.E. and E.S. Parilis, 1986: Sov. Phys.-Tech. Phys. **31**, 15. Sect. 5.2.2

Banouni, M., J. Mischler, M. Négre and N. Benazeth, 1985: Surface Sci. **163**, L720. Sect. 5.5

Baragiola, R.A., 1982: Rad. Effects **61**, 47. Sect. 4.3.5

Baragiola, R.A., E.V. Alonso, O. Auciello, J. Ferrón, G. Lantschner and A. Oliva-Florio, 1978: Phys. Lett. **67A**, 211. Sects. 5.1, 4.3.4, 5.2.1, 5.2.4, (Tab. 5.2)

Baragiola, R.A., E.V. Alonso and A. Oliva-Florio, 1979a: Phys. Rev. B **19**, 121. Sects. 5.1, 3.1.2, 4.1.1, 4.2, 5.2.1, 5.2.4, 5.3, 6, (Tabs. 5.2, 6.1)

Baragiola, R.A., E.V. Alonso, J. Ferrón and A. Oliva-Florio, 1979b: Surface Sci. **90**, 240. Sects. 1, 2, 3.1.2, 4.1.1, 4.2, 5.2.2, 5.3, 6, (Tabs. 1.1, 6.1)

Baragiola, R.A., E.V. Alonso and H.J.L. Raiti, 1982: Phys. Rev. A **25**, 1969. Sect. 3.3

Barat, M., and W. Lichten, 1972: Phys. Rev. A **6**, 211. Sect. 4.1.1

Barnett, C.F. and J.A. Ray, 1972a: Rev. Sci. Instrum. **43**, 218. Sects. 5.1, 6, (Tabs. 5.2, 6.1)

Barnett, C.F. and J.A. Ray, 1972b: Nucl. Fus. **12**, 65. Sect. 6, (Tab. 6.1)

Barnett, C.F., G.E. Evans and P.M. Stier, 1954: Rev. Sci. Instrum. **25**, 1112. Sect. 6

Batrakin, E.N., I.I. Zalyubovkii, V.I. Karas', S.I. Kononenko, V.N. Mel'nik, S.S. Moiseev and V.I. Muratov, 1985: Sov. Phys. JETP **62**, 633. Sect. 5.1, (Tab. 5.3)

Beck, H.P. and R. Langkau, 1975: Z. Naturforsch. **30a**, 981. Sect. 5.1, (Tab. 5.3)

Benazeth, N., 1982: Nucl. Instrum. Meth. **194**, 405. Sects. 1, (Tab. 1.1)

Benazeth, N., 1987: J. Microsc. Spectrosc. Electron. **12**, 235. Sects. 1, 5.4, 5.6, (Tab. 1.1)

Benazeth, N., J. Agusti, C. Benazeth, J. Mischler and L. Viel, 1976: Nucl. Instrum. Meth. **132**, 477. Sect. 6, (Tab. 6.1)

Benazeth, C., N. Benazeth and L. Viel, 1977: Surface Sci. **65**, 165. Sects. 3.3, 6, (Tab. 6.1)

Benazeth, C., N. Benazeth and L. Viel, 1978: Surface Sci. **78**, 625. Sect. 6, (Tab. 6.1)

Benazeth, C., M. Hou, N. Benazeth and C. Mayoral, 1988a: Nucl. Instrum. Meth. Phys. Res. B **30**, 514. Sects. 4.3.5, 6

Benazeth, N., J. Mischler and M. Négre, 1988b: Surface Sci. **205**, 419. Sects. 3.3, 5.1, 5.6, 5.8, (Tab. 5.2)

Benazeth, C., P. Hecquet, C.Mayoral and N. Benazeth, 1989: Rad. Effects Def. Sol. **108**, 227. Sects. 5.1, 5.5, 5.10, (Tab. 5.2)

Bethe, H.A., 1930: Ann. Physik **5**, 325. Sect. 4.2

Bethe, H.A., 1941: Phys. Rev. **59**, 940. Sects. 4.1.1, 4.2, 5.7.2

Bethe H.A., and J. Ashkin, 1960: in *Experimental Nuclear Physics, Vol.I*, ed. by E. Segré, (John Wiley, New York), p. 232. Sects. 2, 4.1.1

Bethge, K. and P. Lexa, 1966: Brit. J. Appl. Phys. **17**, 181. Sects. 5.1, 5.2.3, 5.3, (Tab. 5.2)

Beuhler, R.J., 1983: J. Appl. Phys. **54**, 4118. Sects. 5.1, 5.2.5, 6, (Tabs. 5.2, 6.1)

Beuhler, R.J. and L. Friedman, 1977a: J. Appl. Phys. **48**, 3928. Sects. 4.2, 5.1, (Tab. 5.2)

Beuhler, R.J. and L. Friedman, 1977b: Int. J. Mass Spectrom. Ion Phys. **23**, 81. Sects. 5.1, 5.2.5, 6, (Tabs. 5.2, 6.1)

Beuhler, R.J. and L. Friedman, 1980: Nucl. Instrum. Meth. **170**, 309. Sects. 5.1, 5.2.5, (Tab. 5.3)

Bindi, R., H. Lantéri and P. Rostaing, 1979: J. Electron Spectrosc. Relat. Phenom. **17**, 249. Sect. 5.6

Bindi, R., H. Lantéri and P. Rostaing, 1980: J. Phys. D: Appl. Phys. **13**, 267 and 461. Sect. 5.6

Bindi, R., H. Lantéri and P. Rostaing, 1987: Scanning Microsc. 1, 1475. Sect. 4.3.2

Blaise, G., 1978: in: *Material Characterization Using Ion Beams*, ed. by J.P. Thomas and A. Cachard, (Plenum Press, New York), p. 143. Sect. 3.2.1

Blauth, E.W., W.M. Draeger, J. Kirschner, H. Liebl, N. Müller and E. Taglauer, 1972: J. Vac. Sci. Technol. **8**, 384. Sect. 6, (Tab. 6.1)

Børgesen, P., J. Schou and H. Sørensen, 1980: J. Nucl. Mat. **93/94**, 701. Sects. 5.1, 5.7.2, (Tab. 5.1)

Bøttiger, J.H., H. Wolder Jørgensen and K.B. Winterbon, 1971: Rad. Effects **11**, 133. Sect. 3.2.1

Bohr, N., 1948: Kgl. Danske Videnskab. Selskab., Math.-fys. Medd. **18**(8). Sect. 4.2

Bonnano, A., A. Amoddeo and A. Oliva, 1990: Nucl. Instrum. Meth. Phys. Res. B **46**, 456. Sect. 5.1, (Tab. 5.1)

Boring, J.W., G.E. Strohl and F.R. Woods, 1965: Phys. Rev. **140 A**, 1065. Sect. 4.1.1

Borovsky, J.E. and L. Barraclough, 1989: Nucl. Instrum. Meth. Phys. Res. B **36**, 377. Sects. 5.1, 5.2.1, 5.7.2, 5.9, (Tab. 5.3)

Borovsky, J.E., D.J. McComas and B.L. Barraclough, 1988: Nucl. Instrum. Meth. B **30**, 191. Sects. 5.1, 3.1.2, 5.7.2, (Tab. 5.3)

Bourne, H.C., R.W. Cloud and J.G. Trump, 1955: J. Appl. Phys. **26**, 596. Sect. 6

Brandt, W., and R.A. Ritchie, 1976: Nucl. Instrum. Meth. **132**, 43. Sect. 4.3.4

Brandt, W., A. Ratkowski and R.A. Ritchie, 1974: Phys. Rev. Lett. **33**, 1325. Sect. 4.3.4

Brauer, W. and M. Rösler, 1985: phys. stat. sol. (b) **131**, 177. Sect. 5.2.2

Breunig, J.L., 1972: Phys. Rev. B **6**, 687. Sects. 5.1, 5.8, (Tab. 5.1)

Bruining, H., 1954: *Physics and Applications of Secondary Electron Emission*, (Pergamon Press, London). Sect. 4.2

Brusilovsky, B.A., 1985: Vacuum **35**, 595. Sects. 1, 5.4, (Tab. 1.1)

Brusilovsky, B.A., 1990: Appl. Phys. A **50**, 111. Sects. 1, 4.2, 4.3.5, (Tab. 1.1)

Brusilovsky, B.A. and V.A. Molchanov, 1971: Rad. Effects **8**, 71. Sects. 4.2, 5.1, (Tab. 5.2)

Brusilovsky, B.A. and V.A. Molchanov, 1974: Rad. Effects **23**, 131. Sect. 5.1, (Tab. 5.2)

Brusilovsky, B.A., G.L. Klimenko and V.N. Matveev, 1985: Rad. Effects **90**, 81. Sect. 5.1, (Tab. 5.2)

Burch, D., H. Wieman and W.B. Ingalls, 1973: Phys. Rev. Lett. **30**, 823. Sect. 4.3.3

Burkhard, M.F., H. Rothard and K.-O. Groeneveld, 1988: phys. stat. sol. (b) **147**, 589. Sect. 5.6

Cano, G.L., 1973: J. Appl. Phys. 44, 5293. Sects. 5.1, 5.9, (Tab. 5.1)

Carter, G. and J.S. Colligon, 1968: *Ion Bombardment of Solids*, (Heinemann Educational Books, London). Sects. 1, 4.2, 5.2.3, 5.2.4, (Tab. 1.1)

Cawthron, E.R., 1971: Aust. J. Phys. **24**, 859. Sects. 5.1, 5.2.1, (Tab. 5.1)

Cawthron, E.R., D.L. Cotterell and Sir M. Oliphant, 1969: Proc. Roy. Soc. London A **314**, 39. Sects. 5.1, 5.2.1, 5.2.4, (Tab. 5.2)

Cawthron, E.R., D.L. Cotterell and Sir M. Oliphant, 1970: Proc. Roy. Soc. London A **319**, 435. Sects. 5.1, 5.5, (Tab. 5.2)

Chang, C.C., 1974: in *Characterization of Solid Surfaces*, ed. by P.F. Kane and G.B. Larrabee, (Plenum Press, New York), p. 520. Sect. 3.3

Chanut, Y., J. Martin, , R. Salin and H.O. Moser, 1981: Surface Sci. **106**, 563. Sects. 5.1, 5.2.4, 5.2.5, (Tab. 5.3)

Chapman, B.N., D. Downer and L.J.M. Guimarães, 1974: J. Appl. Phys. **45**, 2115. Sect. 6, (Tab. 6.1)

Chen, H.-L., R. Solarz and G. Erbert, 1983: Appl. Phys. Lett. **42**, 120. Sect. 5.1, (Tab. 5.1)

Chu, W.K., 1978: in *Material Characterization Using Ion Beams*, ed. by J.P. Thomas and A. Cachard, (Plenum Press, New York), p. 3. Sect. 5.2.2

Chung, M.S., and T.E. Everhart, 1974: J. Appl. Phys. **45**, 707. Sect. 4.1.3

Chung, M.S., and T.E. Everhart, 1977: Phys. Rev. B **15**, 4699. Sect. 4.3.2

Clouvas, A., H. Rothard, M. Burkhard, K. Kroneberger, C. Biedermann, J. Kemmler, K.-O. Groeneveld, R. Kirsch, P. Misaelides and A. Katsanos, 1989: Phys. Rev. B **39**, 6316. Sects. 5.2.2, 5.3

Coggiola, M.J., 1986: Nucl. Instrum. Meth. Phys. Res. B **13**, 641. Sect. 5.8

Collins, L.E. and P.T. Stroud, 1967: Brit. J. Appl. Phys. **18**, 1121. Sect. 5.1, (Tab. 5.1)

Cook, N. and R.B. Burtt, 1975: J. Phys. D: Appl. Phys. **8**, 800. Sect. 5.1, (Tab. 5.1)

Dearnaley, G., 1969: Rep. Prog. Phys. **32**, 405. Sect. 1, (Tab. 1.1)

Decoste, R. and B.H. Ripin, 1979: J. Appl. Phys. **50**, 1503. Sects. 3.2.1, 5.1, 5.9, (Tab. 5.2)

Dekker, A.J., 1958: Solid State Physics **6**, 251. Sect. 4.2

Dettmann, K., 1972: in *Proc. Conf. Interaction of Energetic Charged Particles with Solids*, ed. by Goland, A.N., (BNL 50336, Brookhaven National Laboratory, New York), p. 81. Sect. 1, (Tab. 1.1)

Dev, B., 1982: Appl. Surf. Sci. **10**, 240. Sects. 5.1, 5.4, (Tab. 5.1)

Dev, B., 1990: Appl. Surf. Sci. **40**, 319. Sect. 5.1, (Tab. 5.2)

Devienne, F.M., 1967: J. Physique **28**, 602. Sect. 5.1, (Tab. 5.1)

Devooght, J., J.C. Dehaes, A. Dubus, M. Cailler and J.P. Ganachaud, 1991: in *Springer Tracts in Modern Physics*, ed. by G. Höhler, (Springer-Verlag, Berlin), Vol. 122. Sects. 4.1.2, 4.2

Dietz, L.A., 1970: Int. J. Mass Spectrom. Ion Phys. **5**, 11. Sect. 3.2.2

Dietz, L.A. and J.C. Sheffield, 1973: Rev. Sci. Instrum. **44**, 183. Sects. 3.2.2, 5.1, 5.4, (Tab. 5.2)

Dietz, L.A. and J.C. Sheffield, 1975: J. Appl. Phys. **46**, 4361. Sects. 3.2.2, 5.1, 5.2.1, 5.7.2, (Tab. 5.2)

Dorozhkin, A.A. and N.N. Petrov, 1974a: Sov. Phys.-Solid State **16**, 517. Sects. 4.3.4, 5.1, (Tab. 5.1)

Dorozhkin, A.A. and N.N. Petrov, 1974b: Sov. Phys.-Solid State **16**, 611. Sects. 5.1, 5.3, (Tab. 5.1)

Dorozhkin, A.A., A.N. Mishin and N.N. Petrov, 1974a: Bull. Acad. Sci. USSR, Phys. Ser. **38**, 60. Sects. 1, 4.3.1, 5.2.4, (Tab. 1.1)

Dorozhkin, A.A., A.N. Mishin and N.N. Petrov, 1974b: Bull. Acad. Sci. USSR, Phys. Ser. **38**, 249. Sects. 4.3.1, 5.1, (Tab. 5.1)

Dorozhkin, A.A., A.N. Mishin and N.N. Petrov, 1975: Sov. Phys.-Solid State **17**, 1214. Sects. 4.3.1, 5.1, (Tab. 5.1)

Dorozhkin, A.A., A.A. Petrov and N.N. Petrov, 1976a: Bull. Acad. Sci. USSR, Phys. Ser. **40**, 102. Sect. 5.1, (Tab.5.1)

Dorozhkin, A.A., A.A. Petrov and N.N. Petrov, 1976b: Bull. Acad. Sci. USSR, Phys. Ser. **40**, 116. Sects. 3.3, 5.1, 5.6, (Tab. 5.1)

Drepper, F. and J.S. Briggs, 1976: J. Phys. B: At. Mol. Phys. **9**, 2063. Sect. 4.3.3

Dubus, A., J. Devooght and J.C. Dehaes, 1986: Nucl. Instrum. Meth. Phys. Res. B **13**, 623. Sects. 4.1.2, 4.2

Eckstein, W. and J.P. Biersack, 1983: Z. Physik A **310**, 1. Sect. 3.2.1

Erginsoy, C., 1967: in *Proc. Conf. Solid State Research with Accelerators*, ed. by A.N. Goland, (Brookhaven report, BNL-50083, (C-52)), 30. Sect. 4.3.2

Ertl, K. and R. Behrisch, 1984: in *Physics of Plasma-Wall Interactions in Controlled Fusion*, ed. by D.E. Post and R. Behrisch, (Plenum Press, New York), p. 515. Sect. 1, (Tab. 1.1)

Evdokimov, I.N. and V.A. Molchanov, 1968: Phys. Lett. **26A**, 636. Sect. 5.1, (Tab. 5.2)

Evdokimov, I.N., E.S. Mashkova, V.A. Molchanov and D.D. Odintsov, 1967a: phys. stat. sol. **19**, 407. Sects. 5.1, 5.4, (Tab. 5.2)

Evdokimov, I.N., V.A. Molchanov, D.D. Odintsov and V.M. Chicherov, 1967b: Sov. Phys.-Solid State **8**, 2348–2352. Sect. 5.1, (Tab. 5.2)

Evdokimov, I.N., E.S.Mashkova and V.A. Molchanov, 1967c: Sov. Phys.-Solid State **9**, 1434–1435. Sect. 6, (Tab. 6.1)

Even, U., P.J. de Lange, H.T. Jonlman and J. Kommandeur, 1986: Phys. Rev. Lett. **56**, 965. Sect. 5.2.5

Everhart, T.E., N. Saeki, R. Shimizu and T. Koshikawa, 1976: J. Appl. Phys. **47**, 2941. Sects. 4.3.2, 5.6

Ewing, R.I., 1968: Phys. Rev. **166**, 324. Sects. 5.1, 5.2.1, (Tab. 5.2)

Faizan-Ul-Haq and M.A. Chaudhry, 1976: Lett. Nuovo Cim. **16**, 574. Sect. 5.1, (Tab. 5.1)

Fano, U., and W. Lichten, 1965: Phys. Rev. Lett. **14**, 627. Sects. 4.1.1

Feder, R., 1985: *Polarized Electrons in Surface Physics*, (World Scientific, Singapur). Sect. 5.11

Fehn, U., 1976: Int. J. Mass Spectrom. Ion Phys. **21**, 1. Sects. 5.1, 5.2.1, 5.3, (Tab. 5.1)

Feijen, H.H.W., L.K. Verhey, A.L. Boers and E.P.T.M. Suurmeyer, 1973: J. Phys. E: Sci. Instrum. **6**, 1174. Sect. 6, (Tab. 6.1)

Ferguson, M.M. and W.O. Hofer, 1989: Rad. Effects Def. Sol. **109**, 273. Sects. 3.2.2, 5.1, 5.3, (Tab. 5.1)

de Ferrariis, L.F. and R.A. Baragiola, 1986: Phys. Rev. A **33**, 4449. Sects. 5.1, 5.6, (Tab. 5.2)

Ferrón, J., A.V. Alonso, R.A. Baragiola and A. Oliva-Florio, 1981a: Phys. Rev. B **24**, 4412. Sects. 4.3.1, 5.1, 5.4, (Tab. 5.2)

Ferrón, J., E.V. Alonso, R.A. Baragiola and A. Oliva-Florio, 1981b: J. Phys. D: Appl. Phys. **14**, 1707. Sects. 3.1.1, 4.2, 4.3.1, 5.1, 5.2.1, 5.7.2, (Tab. 5.2)

Ferrón, J., E.V. Alonso, R.A. Baragiola and A. Oliva-Florio, 1982: Surface Sci. **120**, 427. Sects. 5.1, 5.8, (Tab. 5.2)

Firsov, O.B., 1958: Sov. Phys.-JETP **7**, 308. Sect. 4.2

Firsov, O.B., 1959: Sov. Phys.-JETP **9**, 1517. Sect. 4.2

Fitzwilson, R.L. and E.W. Thomas, 1971: Rev. Sci. Instrum. **42**, 1864. Sect. 6, (Tab. 6.1)

Folkmann, F., C. Gaarde, T. Huus and K. Kemp, 1974: Nucl. Instrum. Meth. **116**, 487. Sect. 6, (Tab. 6.1)

Fontbonne, J.-P., N. Colombie and B. Fagot, 1970: C. R. Acad. Sc. Paris B **270**, 1573. Sect. 5.1, (Tab. 5.2)

Foti, G., R. Potenza and A. Triglia, 1974: Lett. Nouvo Cim. **11**, 659. Sect. 5.1, (Tab. 5.3)

Frischkorn, J. and K.-O. Groeneveld, 1983: Phys. Scripta **T6**, 89. Sect. 5.3

Frischkorn, J., K.-O. Groeneveld, D. Hofmann, P. Koschar, R. Latz and J. Schader, 1983: Nucl. Instrum. Meth. **214**, 123. Sect. 5.3

Füchtbauer, C., 1906: Physik. Z. **7**, 153 and 748. Sects. 5.6, 6

Garcia, J.D., 1970: Phys. Rev. A **1**, 280. Sect. 4.3.6

Gaworzewski, P., K.H. Krebs and M. Mai, 1972/73: Int. J. Mass Spectrom. Ion Phys. **10**, 425. Sects. 3.3, 5.1, 5.3, 5.7.1, 5.7.2, (Tab. 5.1)

Gaworzewski, P., K.H. Krebs and M. Mai, 1974: Int. J. Mass Spectrom. Ion Phys. **13**, 99. Sects. 5.1, 5.3, 5.7.1, (Tab. 5.1)

Geiger, H., 1927: in *Handbuch der Physik, Vol. 24*, ed. by H. Geiger, (Springer-Verlag, Berlin), p. 171. Sects. 1, 2, (Tab. 1.1)

von Gemmingen, U., 1982: Surface Sci. **120**, 334. Sect. 5.1, (Tab. 5.1)

Gerjuoy, E., 1966: Phys. Rev. **148**, 54. Sect. 4.3.6

Giber, J., I. Nagy and J. László, 1984: Nucl. Instrum. Meth. Phys. Res. B **2**, 135. Sect. 4.3.1

van Gorkum, M. and R.E. Glick, 1970: Int. J. Mass Spectrom. Ion Phys **4**, 203. Sects. 5.1, 6, (Tabs. 5.1, 6.1)

Gorodetzky, M.S., A.M. Bergdolt, A. Chevalier, M. Bres and R. Armbruster, 1963: J. Physique **24**, 374. Sects. 5.1, 5.4, 5.6, (Tab. 5.3)

Gosh, S.N. and S.P. Khare, 1962: Phys. Rev. **125**, 1254. Sects. 4.3.4, 5.2.3, 5.2.4

Gosh, S.N. and S.P. Khare, 1963: Phys. Rev. **129**, 1638. Sects. 4.3.4, 5.2.3, 5.2.4

Gosh, S.N. and B.K. Majumdar, 1980: Indian J. Phys. **54A**, 246. Sect. 5.2.3

Groeneveld, K.-O., 1988: Nucl. Tracks Radiat. Meas. **15**, 51. Sect. 6, (Tab. 6.1)

Hachenberg, O., and W. Brauer, 1959: Adv. Electron. Electron Phys. **11**, 413. Sects. 3.3, 4.1.3

Hagstrum, H.D., 1954: Phys. Rev. **96**, 336. Sect. 1

Hagstrum, H.D., 1978: in *Electron and Ion Spectroscopy of Solids*, ed. by L. Fiermans, J. Vennik and W. Dekeyser, (Plenum Press, New York), p. 273. Sects. 1, 2

Hanak, J.J. and J.P. Pellicane, 1976: J. Vac. Sci. Technol. **13**, 406. Sect. 6, (Tab. 6.1)

Hasegawa, M., K. Kimura, Y. Fujii, M. Suzuki, Y. Susuki and M. Mannami, 1988: Nucl. Instrum. Meth. Phys. Res. B **33**, 334. Sects. 5.1, 5.4, 5.7.1, (Tab. 5.3)

Hasselkamp, D., 1985: Habilitationsschrift, Giessen, FRG. Sects. 1, 4.2, 4.3.2, 5.2.2, 5.7.2, (Tab. 1.1)

Hasselkamp, D., 1988: Comments At. Mol. Phys. **21**, 241. Sects. 1, 5.3, (Tab. 1.1)

Hasselkamp, D., A. Scharmann, 1982a: Surface Sci. **119**, L388. Sects. 5.1, 5.6, (Tab. 5.3)

Hasselkamp, D., A. Scharmann, 1982b: Vak.-Tech. **31**, 242. Sects. 3.1.2, 3.3, 5.1, 5.7.2, 6, (Tabs. 5.3, 6.1)

Hasselkamp, D. and A. Scharmann, 1983a: Vak.-Tech. **32**, 9. Sects. 5.1, 5.6, (Tab. 5.3)

Hasselkamp, D. and A. Scharmann, 1983b: Phys. Lett. **96A**, 259. Sect. 5.1, (Tab. 5.2)

Hasselkamp, D. and A. Scharmann, 1983c: phys. stat. sol. (a) **79**, K197. Sects. 5.1, 5.2.1, 5.2.2, 5.2.4, (Tab. 5.3)

Hasselkamp, D., A. Scharmann and N. Stiller, 1980: Nucl. Instrum. Meth **168**, 579. Sects. 3.3, 5.1, 5.8, (Tab. 5.3)

Hasselkamp, D., K.G. Lang, A. Scharmann and N. Stiller, 1981: Nucl. Instrum. Meth. **180**, 349. Sects. 3.3, 5.1, 5.2.1, 5.2.2, 5.2.4, (Tab. 5.3)

Hasselkamp, D., S. Hippler and A. Scharmann, 1984: Nucl. Instrum. Meth. Phys. Res. B **2**, 475. Sects. 4.3.4, 5.1, 5.2.4, 5.6, (Tab. 5.3)

Hasselkamp, D., S. Hippler and A. Scharmann, 1987a: Nucl. Instrum. Meth. Phys. Res. B **18**, 561. Sects. 3.1.2, 5.1, 5.6, 6.7.1, 5.7.2, (Tab. 5.3)

Hasselkamp, D., S. Hippler, A. Scharmann and K.-H. Schartner, 1987b: Z. Phys. D **6**, 269. Sects. 4.3.4, 5.1, 5.2.4, 5.6, (Tab. 5.3)

Hasted, J.B., 1964: *Physics of Atomic Collisions*, (Butterworths, London), Chap. 3.12. Sects. 3.2.1, 6

Heimann, P.A. and J. Blakeslee, 1986: J. Electrochem. Soc. (USA) **133**, 779. Sect. 6, (Tab. 6.1)

Hennequin, J.F. and P. Viaris de Lesegno, 1980: in *Proc. Conf. Physics of Ionized Gases*, ed. by B. Cobić, (Boris Kidrich Institute, Beograd), p. 341. Sect. 4.3.5

Hepworth, J.K., 1970: J. Phys. D: Appl. Phys. **3**, 1475. Sect. 5.1, (Tab. 5.2)

Hill, A.G., W.W. Buechner, J.S. Clark and J.B. Fisk, 1939: Phys. Rev. **55**, 463. Sect. 5.2.4, 6,

Hippler, S., 1988: Thesis, Giessen, FRG. Sects. 5.2.1, 5.3, 5.6, 5.7.1

Hippler, S., D. Hasselkamp and A. Scharmann, 1988: Nucl. Instrum. Meth. Phys. Res. B **34**, 518. Sects. 5.1, 5.2.2, 5.3, (Tab. 5.3)

Hird, B., C. Pepin and G. Kelly, 1984: J. Appl. Phys. **56**, 3304. Sects. 5.1, 5.2.3, (Tab. 5.2)

Hofer, W.O., 1980: Nucl. Instrum. Meth. **170**, 275. Sects. 5.1, 5.2.5, (Tab. 5.1)

Hofer, W.O., 1987: J. Vac. Sci. Technol. A **5**, 2213. Sect. 6, (Tab. 6.1)

Hofer, W.O., 1989: Report KFA Jülich, FRG, (Jül-2317, ISSN 0366-0885, Nov. 1989), and Scanning Microsc., Suppl. 4 (1990), 265. Sects. 1, 3.2.2, 4.1.1, 5.4, 5.5, (Tab. 1.1)

Holmén, G. and P. Högberg, 1972: Rad. Effects **12**, 77. Sects. 5.1, 5.10, 6, (Tabs. 5.2, 6.1)

Holmén, G., P. Högberg and A. Burén, 1975a: Rad. Effects **24**, 39. Sects. 5.1, 5.10, 6, (Tabs. 5.2, 6.1)

Holmén, G, , S. Peterström and A. Burén, 1975b: Rad. Effects **24**, 45. Sects. 5.1, 5.10, 6, (Tabs. 5.2, 6.1)

Holmén, G., A. Burén and P. Högberg, 1975c: Rad. Effects **24**, 51. Sects. 5.1, 5.10, 6, (Tabs. 5.2, 6.1)

Holmén, G., B. Svensson, J. Schou and P. Sigmund, 1979: Phys. Rev. B **20**, 2247. Sects. 4.2, 4.3.1, 5.2.2

Holmén, G., B. Svensson and A. Burén, 1981: Nucl. Instrum. Meth. **185**, 523. Sects. 3.1.2, 4.2, 4.3.1, 5.1, 5.2.1, 5.2.2, (Tab. 5.3)

Inokuti, M., 1971: Rev. Mod. Phys. **43**, 297. Sect. 5.2.2

Inokuti, M., 1983: in *Applied Atomic Collision Physics, Vol. 4*, ed. by S. Datz, (Academic Press, New York), p. 179. Sect. 6

Janni, J.F., 1982: At. Data Nucl. Data Tables **27**, 341. Sect. 5.2.2

Jonkers, J.L.H., 1957: Philips Res. Rep. **12**, 249. Sect. 4.1.3

Kaminski, M., 1965: *Atomic and Ionic Impact Phenomena on Solid Surfaces*, (Springer-Verlag, Berlin). Sect. 1, (Tab. 1.1)

Kim, Y. and K. Cheng, 1980: Phys. Rev. A **22**, 61. Sect. 5.9

Kirschner, J., 1985: *Polarized Electrons at Surfaces*, Springer Tracts in Modern Physics, Vol. 123 (Springer, Berlin Heidelberg). Sect. 5.11

Kirschner, J., K. Koike and H.P. Oepen, 1987: Phys. Rev. Lett. **59**, 2099. Sect. 5.11

Kishinevskii, L.M., 1973: Rad. Effects **19**, 23. Sect. 2

Kishinevskii, L.M. and E.S. Parilis, 1962: Bull. Acad. Sci. USSR, Phys. Ser. **26**, 1432. Sect. 4.2

Kislyakov, A.I., J. Stöckel and K. Jalubka, 1976: Sov. Phys.-Tech. Phys. **20**, 986. Sect. 6, (Tab. 6.1)

Klein, H.J., 1965: Z. Physik **188**, 78. Sects. 3.4, 5.5

Kneis, H., B. Martin, R. Nobeling, B. Povh and K. Traxel, 1982: Nucl. Instrum. Meth. **197**, 79. Sect. 6, (Tab. 6.1)

König, W., K.H. Krebs and S. Rogaschewski, 1975: Int. J. Mass Spectrom. Ion Phys. **16**, 243. Sects. 3.1, 3.2.1, 5.1, 5.7.2, (Tab. 5.1)

Koschar, P., K. Kronebergr, A. Clouvas, M. Burkhard, W. Meckbach, O. Heil, J. Kemmler, H. Rothard, K.-O. Groeneveld, R. Schramm, H.-D. Betz, 1989: Phys . Rev. A **40**, 3632. Sect. 4.2

Koyama, A., E. Yagi and H. Sakairi, 1976: Jpn. J. Appl. Phys. **15**, 1811. Sects. 5.1, 5.2.2, 5.9, (Tab. 5.3)

Koyama, A., T. Shikata and H. Sakairi, 1981: Jpn. J. Appl. Phys. **20**, 65. Sects. 5.1, 5.2.1, 5.2.2, (Tab. 5.3)

Koyama, A., T. Shikata, H. Sakairi and E. Yagi, 1982a: Jpn. J. Appl. Phys. **21**, 586. Sects. 5.1, 5.3, 5.9, (Tab. 5.3)

Koyama, A., T. Shikata, H. Sakairi and E. Yagi, 1982b: Jpn. J. Appl. Phys. **21**, 1216. Sect. 5.1, (Tab. 5.3)

Koyama, A., O. Benka, Y. Sasa and M. Uda, 1986a: Nucl. Instrum. Meth. Phys. Res. B **13**, 637. Sects. 5.1, 5.6, 5.9, (Tab. 5.3)

Koyama, A., O. Benka, Y. Sasa and M. Uda, 1986b: Phys. Rev. B **34**, 8150. Sects. 5.1, 5.9, (Tab. 5.3)

Koyama, A., O. Benka, Y. Sasa, H. Ishikawa and M. Uda, 1987: Phys. Rev. A **36**, 4535. Sects. 5.1, 5.6, (Tab. 5.3)

Koyama, A., H. Ishikawa, Y. Sasa, O. Benka and M. Uda, 1988a: Nucl. Instrum. Meth. Phys. Res. B **33**, 338. Sects. 5.1, 5.6, 5.9, (Tab. 5.3)

Koyama, A., H. Ishikawa, , Y. Sasa, O. Benka and M. Uda, 1988b: Nucl. Instrum. Meth. Phys. Res. B **33**, 341. Sects. 5.1, 5.9, (Tab. 5.3)

Krebs, K.H., 1968: Fortschr. d. Physik **16**, 419. Sects. 1, 3.1.1, 3.2.1, 3.2.2, 3.3, 4.2, 5.1, 5.2.1, 5.7.2, 5.10, 6, (Tab. 1.1)

Krebs, K.H., 1976: in *Proc. 8th Int. Summer School on the Physics of Ionized Gases, Dubrovnik*, (Univ. Ljubljana 1976), p. 379. Sects. 1, 4.2, (Tab. 1.1)

Krebs, K.H., 1980: in *Proc. Symposium on Sputtering, Perchtoldsdorf/Wien*, ed. by P. Varga, G. Betz and F.P. Viehböck, (Techn. Univ. Wien), p. 642. Sects. 1, 4.2, (Tab. 1.1)

Krebs, K.H., 1983a: Vacuum **33**, 7. Sect. 6, (Tab. 6.1)

Krebs, K.H., 1983b: Vacuum **33**, 555. Sect. 1, 5.2.4, (Tab. 1.1)

Krebs, K.H. and S. Rogaschewski, 1976: Wiss. Z. Humboldt-Univ. Berlin, Math.-Nat. R. **25**, 385. Sects. 1, 4.2, (Tab. 1.1)

Kroneberger, K., A. Clouvas, G. Schüssler, P. Koschar, J. Kemmler, H. Rothard, C. Biedermann, O. Heil, M. Burkhard and K.-O. Groeneveld, 1988: Nucl. Instrum. Meth. Phys. Res. B **29**, 621. Sect. 5.2.4

Kroneberger, K., H. Rothard, M. Burkhard, J. Kemmler, P. Koschar, O. Heil, C. Biedermann, S. Lencinas, N. Keller, P. Lorenzen, D. Hofmann, A. Clouvas, E. Veje and K.-O. Groeneveld, 1989: J. Physique Coll. C2 **50**, 99. Sect. 5.2.4

Kronenberg, S., K. Nilson and M. Basso, 1961: Phys. Rev. **124**, 1709. Sects. 5.1, 5.6, (Tab. 5.3)

Lai, S.Y., D. Briggs, A. Brown and J.C. Vickerman, 1986: Surf. Interface Anal. **8**, 93. Sects. 1, 6, (Tabs. 1.1, 6.1)

Lakits, G., 1989: Thesis, Technische Universität Wien, Austria. Sect. 3.2.2

Lakits, G. and H. Winter, 1990: Nucl. Instrum. Meth. Phys. Res. B **48**, 597. Sects. 3.2.2, 5.1, 5.2.2, (Tab. 5.1)

Lakits, G., F. Aumayr and H. Winter, 1989a: Europhys. Lett. **10**, 679. Sects. 3.2.2, 4.3.4, 5.1, 5.2.1, 5.2.3, 5.2.4, (Tab. 5.1)

Lakits, G., F. Aumayr and H. Winter, 1989b: Rev. Sci. Instrum. **60**, 3151. Sects. 3.2.2, 5.1, 5.2.1, 5.2.4, (Tab. 5.1)

Lakits, G., F. Aumayr and H. Winter, 1989c: Phys. Lett. A **139**, 395. Sects. 3.2.2, 5.1, (Tab. 5.1)

Lakits, G., F. Aumayr and H. Winter, 1989d: J. Physique Coll. C1 **50**, 533. Sect. 5.1, (Tab. 5.1)

Lakits, G., A. Arnau and H. Winter, 1990: Phys. Rev. B **42**, 15. Sects. 5.1, 5.2.3, (Tab. 5.1)

Lao, R.C., R. Sander and R.F. Pottie, 1972/1973: Int. J. Mass Spectrom. Ion Phys. **10**, 309. Sects. 5.1, 6, (Tabs. 5.1, 6.1)

Large, L.N. and W.S. Whitlock, 1962: Proc. Phys. Soc. **79**, 148. Sects. 5.2.1, 5.3

Layton, J.K., 1973: J. Chem. Phys. **59**, 5744. Sects. 5.1, 5.2.3, (Tab. 5.1)

Lebedev, S.Ya. and N.M. Omel'yanovskaya, 1974: Rad. Effects **22**, 135. Sects. 5.1, 5.10, (Tab. 5.2)

Lebedev, S.Ya., N.M. Omel'yanovskaya and V.I. Krotov, 1969: Sov. Phys.-Solid State **11**, 1294. Sects. 5.1, 5.6, (Tab. 5.2)

Levi-Setti, R., 1983: Scanning Electron Microsc. **1**, 1. Sect. 6, (Tab. 6.1)

Levi-Setti, R., G. Crow and Y.L. Wang, 1985: Scanning Electron Microsc. **2**, 535. Sect. 6, (Tab. 6.1)

Levi-Setti, R., J.M. Chabala and Y.L. Wang, 1988: Ultramicrosc. **24**, 97. Sect. 6, (Tab. 6.1)

Lewis, M.A., and D.A. Glocker, 1989: J. Vac. Sci. Technol. A **7**, 1019. Sect. 6, (Tab. 6.1)

Lindhardt, J., and A. Winter, 1964: Kgl. Danske Videnskab. Selskab, Math.-Fys. Medd. **34 (4)**. Sect. 4.2

Linford, L.H., 1935: Phys. Rev. **47**, 279. Sect. 5.2.1

Little, P.F., 1956: in *Handbuch der Physik, Vol. 21*, ed. by S. Flügge, (Springer-Verlag, Berlin), p. 574. Sects. 1, 6, (Tab. 1.1)

Losch, W.H.P., 1970: phys. stat. sol. (a) **2**, 123. Sects. 3.3, 5.1, 5.2.1, 5.5, (Tab. 5.1)

Lukirsky, P., 1924: Z. Physik **22**, 351. Sect. 3.3

Märk, T.D., 1977: Z. Naturforsch. **32a**, 1559. Sect. 5.1, (Tab. 5.1)

Mahadevan, P., G. Magnuson, J.K. Layton and C.E. Carlston, 1965: Phys. Rev. A **140**, 1407. Sect. 5.2.3

Makarov, V.V., and N.N. Petrov, 1981: Sov. Phys.-Solid State **23**, 1028. Sect. 5.3

Massey, H.S.W. and E.H.S. Burhop, 1952: *Electronic and Ionic Impact Phenomena*, (At the Clarendon Press, Oxford). Sects. 1, 4.1.1, 5.3, 6, (Tab. 1.1)

Matteson, S. and M.A. Nicolet, 1979: Nucl. Instrum. Meth. **160**, 301. Sect. 6, (Tab. 6.1)

Matthew, J.A.D., 1983: Phys. Scripta **T6**, 79. Sect. 4.3.5

McCracken, G.M., 1975: Rep. Progr. Phys. **38**, 241. Sect. 1, (Tab. 1.1)

McCracken, G.M. and P.E. Stott, 1979: Nucl. Fus. **19**, 889. Sect. 6, (Tab. 6.1)

McDonald, R.J., 1970: Adv. Phys. **19**, 514. Sect. 1, (Tab. 1.1)

Medved, D.B. and Y.E. Strausser, 1965: in *Advances in Electronics and Electron Physics*, ed. by Marton, L., (Academic Press, New York), p. 101. Sects. 1, 4.2, (Tab. 1.1)

Melngailis, J., 1987: J. Vac. Sci. Technol. B **5**, 469. Sect. 6, (Tab. 6.1)

Mischler, J., 1987: Rad. Effects **105**, 133. Sects. 3.1.2, 5.5

Mischler, J., N. Benazeth, M.Négre and C. Benazeth, 1984: Surface Sci. **136**, 532. Sects. 3.4, 5.5

Mischler, J., M. Banouni, C. Benazeth, M. Négre and N. Benazeth, 1986: Rad. Effects **97**, 1. Sects. 3.4, 5.5

Möllenstedt, G. and F. Lenz, 1963: Adv. Electronics Electron Phys. **18**, 251. Sects. 1, 6, (Tabs. 1.1, 6.1)

Monnin, M., 1975: in *Radiation Damage Processes in Materials*, ed. by C.H.S. Dupuy, (Noordhoff, Leyden), p. 361. Sect. 6, (Tab. 6.1)

Morita, T. and H. Hashimoto, 1988: Jap. J. Appl. Phys. **27**, 1759. Sect. 6, (Tab. 6.1)

Morita, K., H. Akimune and T. Suita, 1966: Jpn. J. Appl. Phys. **5**, 511. Sects. 5.1, 5.2.3, (Tab. 5.2)

Moshammer, R. and R. Matthäus, 1989: J. Physique Coll. C2 **50**, 111. Sects. 5.1, 5.2.5, 5.7.2, 6, (Tabs. 5.1, 6.1)

Musket, R.G., 1975: J. Vac. Sci. Technol. **12**, 444. Sects. 4.3.5, 5.1, 5.2.1, (Tab. 5.3)

Musket, R.G., and W. Bauer, 1972: Appl. Phys. Lett. **20**, 455. Sect. 4.3.5

Musket, R.G., W. McLean, C.A. Columenares, D.M. Makowiecki and W.J. Siekhaus, 1982: Appl. Surf. Sci. **10**, 143. Sect. 3.1.1

Nagy, I., J. László and J. Giber, 1983: Appl. Phys. A **31**, 153. Sect. 4.3.1

Needham, P.B., T.J. Driscoll, C.J. Powell and R.J. Stein, 1977: Appl. Phys. Lett. **30**, 357. Sect. 6, (Tab. 6.1)

Nikolaev, E.N., G.D. Tantsyrev and V.A. Saraev, 1978: Sov. Phys.-Tech. Phys. **23**, 241. Sects. 5.1, 5.2.5, 6, (Tabs. 5.1, 6.1)

Northcliffe, L.C., and R.F. Schilling, 1970: Nucl. Data Tabl. A **7**, 233. Sect. 5.2.2

Oda, N. and J.T. Lyman, 1967: Rad. Res. Suppl. **7**, 20. Sect. 6, (Tab. 6.1)

Oda, N., F. Nishimura, Y. Yamazaki and S. Tsurubuchi, 1980: Nucl. Instrum. Meth. **170**, 571. Sect. 5.6

Paetow, H. and W. Walcher, 1938: Z. Physik **110**, 69. Sect. 5.8

Palmberg, P.W., 1967: J. Appl. Phys. **38**, 2137. Sect. 3.3

Parikh, M. and R. Shimizu, 1976: Appl. Phys. Lett. **29**, 516. Sect. 5.2.2

Parilis, E.S., 1962: Radio Eng. Electron. Phys. **7**, 1839. Sect. 1, (Tab. 1.1)

Parilis, E.S. and L.M. Kishinevskii, 1960: Sov.Phys.-Solid State **3**, 885. Sects. 4.1.1, 4.2, 5.2.1, 5.3

Penn, D.R., 1987: Phys. Rev. B **35**, 482. Sect. 4.1.2

Perdrix, M., S. Paletto, R. Goutte and C. Guillaud, 1968: Brit. J. Appl. Phys. **1**, 1517. Sects. 5.1, 5.4, 5.10, (Tab. 5.2)

Perdrix, M., S. Paletto, R. Goutte and C. Guillard, 1969a: Brit. J. Appl. Phys. **2**, 441. Sects. 5.1, 5.2.1, 5.2.3, 5.4, 5.8, (Tab. 5.2)

Perdrix, M., S. Paletto, R. Goutte and C. Guillaud, 1969b: Phys. Lett. **28A**, 534. Sects. 5.1, 5.9, (Tab. 5.2)

Perdrix, M., J.C. Baboux, R. Goutte and C. Guillaud, 1970: J. Phys. D: Appl. Phys. **3**, 594. Sects. 5.1, 5.8, (Tab. 5.1)

Petrov, N.N., 1960: Sov. Phys.-Solid State **2**, 1182. Sects. 1, 4.1.1, (Tab. 1.1)

Petrov, N.N., 1974: in *Physics of Ionized Gases*, ed. by Vujnovic, V., (Zagreb), p. 533. Sect. 1, (Tab. 1.1)

Pillon, J., D. Roptin and M. Cailler, 1976: Surface Sci. **59**, 741. Sect. 5.6

Ploch, W., 1951: Z. Physik **130**, 174. Sects. 4.1.1, 5.2.1, 5.3

Pottie, R.F., D.L. Cocke and K.A. Gingerich, 1973: Int. J. Mass Spectrom. Ion Phys. **11**, 41. Sects. 5.1, 6, (Tabs. 5.1, 6.1)

Powell, C.J. and R.J. Stein, 1977: Phys. Rev. B **16**, 1370. Sect. 6, (Tab. 6.1)

Powell, C.J., N.E. Erickson and T. Jach, 1982a: J. Vac. Sci. Technol. **20**, 625. Sect. 3.3

Powell, C.J., N.E. Erickson and T.E. Madey, 1982b: J. Electron Spectrosc. Relat. Phenom. **25**, 87. Sect. 3.3

Pradel, P., F. Roussel, A.S. Schlachter, G. Spiess and A. Valence, 1974: Phys. Rev. A **10**, 797. Sect. 6, (Tab. 6.1)

Prewett, P.D., 1984: Vacuum **34**, 931. Sect. 6, (Tab. 6.1)

Probst, F.M. and E. Lüscher, 1963: Rev. Sci. Instrum. **34**, 574. Sect. 3.4

Raether, H., 1980: *Excitation of Plasmons and Interband Transitions by Electrons*, Springer Tracts in Modern Physics, Vol. 88 (Springer-Verlag, Berlin). Sect. 4.3.2

Ray, J.A. and C.F. Barnett, 1971: J. Appl. Phys. **42**, 3260. Sect. 5.1, (Tab. 5.1)

Ray, J.A., C.F. Barnett and B. van Zyl, 1979: J. Appl. Phys. **50**, 6516. Sect. 6, (Tab. 6.1)

Rinn, K., A. Müller, H. Eichenauer and E. Salzborn, 1982: Rev. Sci. Instrum. **53**, 829. Sect. 6, (Tab. 6.1)

Rivière, J.C., 1983: in *Practical Surface Analysis*, ed. by D. Briggs and M.P. Seah, (John Wiley, New York), p. 70. Sect. 3.3

Rogaschewski, S. and H. Düsterhöft, 1976: phys. stat. sol. (b) **75**, K173. Sects. 5.1, 5.3, (Tab. 5.2)

Rösler, M., and W. Brauer, 1981: phys. stat. sol. (b), **104**, 161 and 575. Sect. 4.3.2

Rösler, M. and W. Brauer, 1984: phys. stat. sol. (b) **126**, 629. Sects. 4.1.1, 4.1.2, 4.2, 4.3.2, 5.6

Rösler, M. and W. Brauer, 1988: phys. stat. sol. (b) **148**, 213. Sects. 4.1.1, 4.1.2, 4.2, 4.3.2, 5.6

Rösler, M. and W. Brauer, 1989: phys. stat. sol. (b) **156**, K85. Sect. 5.2.2

Rösler, M., and W. Brauer, 1991: in *Springer Tracts in Modern Physics*, ed. by G. Höhler, (Springer-Verlag, Berlin), Vol. 122. Sects. 4.1.1, 4.1.2, 4.2, 4.3.2, 5.2.2, 5.6

Rothard, H., P. Lorenzen, N. Keller, O. Heil, D. Hofmann, J. Kemmler, K. Kroneberger, S. Lencinas and K.-O. Groeneveld, 1988: Phys. Rev. B **38**, 9224. Sects. 5.1, 5.10, (Tab. 5.3)

Rothard, H., K. Kroneberger, M. Burkhard, J. Kemmler, P. Koschar, O. Heil, C. Biedermann, S. Lencinas, N. Keller, P. Lorenzen, D. Hofmann, A. Clouvas, K.-O. Groeneveld and E. Veje, 1989: Rad. Effects Def. Sol. **109**, 281. Sects. 5.2.2, 5.3

Rothard, H., K. Kroneberger, A. Clouvas, E. Veje, P. Lorenzen, N. Keller, J. Kemmler, W. Meckbach, K.-O. Groeneveld, 1990a: Phys. Rev. A **41**, 2521. Sects. 5.2.2, 5.3

Rothard, H., K. Kroneberger, E.Veje, A. Clouvas, J. Kemmler, P. Koschar, N. Keller, S. Lencinas, P. Lorenzen, O. Heil, D. Hofmann, K.-O. Groeneveld, 1990b: Phys. Rev. B **41**, 3959. Sects. 4.3.4, 5.2.4, 6, (Tab. 6.1)

Roy, D., and J.D. Carette, 1977: in *Electron Spectroscopy for Surface Analysis*, ed. by H. Ibach, (Springer-Verlag, Berlin), p. 13. Sect. 3.3

Rudat, M.A. and G.H. Morrison, 1978: Int. J. Mass Spectrom. Ion Phys. **27**, 249. Sects. 5.1, 6, (Tabs. 5.2, 6.1)

Rudd, M.E., and J.H. Macek, 1972: Case Studies in Atomic Physics **3**, 47. Sect. 4.1.1

Rudd, M.E., C.A. Sautter and C.L. Bailey, 1966: Phys. Rev. **151**, 20. Sect. 4.1.1

Rudd, M.E., J.S. Risley, J. Fryar and R.G. Rolfes, 1980: Phys. Rev. A **21**, 506. Sect. 4.3.3

Rüchardt, E., 1927: in *Handbuch der Physik, Vol. 24*, ed. by H. Geiger and K. Scheel, (Springer-Verlag, Berlin), p. 105. Sects. 1, 2, (Tab. 1.1)

Rutherford, E., 1905: Phil. Mag. **10**, 193. Sect. 1

Salow, H., 1940: Phys. Z. **41**, 434. Sect. 4.2

Schackert, P., 1966: Z. Physik **197**, 32. Sects. 5.1, 5.2.3, (Tab. 5.1)

Schou, J., 1980: Phys. Rev. B **22**, 2141. Sects. 3.3, 4.1.2, 4.1.3, 4.2, 4.3.1, 5.2.2, 5.6

Schou, J., 1988: Scanning Microsc. **2**, 607. Sects. 1, 4.1.3, 4.2, 5.2.2, 5.6, 5.6, (Tab. 1.1)

Schram, B.L., A.J.H. Boerboom, W. Kleine and J. Kistemaker, 1966: Physica **32**, 749. Sects. 5.1, 5.9, (Tab. 5.2)

Seah, M.P., 1989: in *Methods of Surface Analysis*, ed. by J.M. Walls, (Cambridge University Press, Cambridge), p. 57. Sect. 3.3

Seah, M.P., and W.A. Dench, 1979: Surf. Interface Anal. **1**, 2. Sects. 4.1.2, 4.2

Seiler, H., 1967: Z. Angew. Phys. **22**, 249. Sect. 4.1.2

Sidenius, G. and T. Lenskjaer, 1976: Nucl. Instrum. Meth. **132**, 673. Sect. 3.2.1

Sigmund, P., 1975: in *Radiation Damage Processes in Materials*, ed. by C.H.S. Dupuy, (Noordhoff, Leyden), p. 1. Sect. 5.2.2

Sigmund, P. and S. Tougaard, 1981: in *Inelastic Particle-Surface Collisions*, ed. by

E. Taglauer and W. Heiland, (Springer-Verlag, Berlin), p. 2. Sects. 1, 4.1.1, 4.1.2, 4.2, 4.3.1, 4.3.5, 5.2.1, (Tab. 1.1)

Simon, R., N. Colombie, P. van Chuong and L. Dandurand, 1962: Comptes Rendus **255**, 1217. Sect. 5.10

Simpson, J.A., 1961: Rev. Sci. Instrum. **32**, 1283. Sect. 3.3

Sørensen, H., 1976: Appl. Phys. **9**, 321. Sects. 3.2.1, 5.1, 5.7.2, (Tab. 5.1)

Sørensen, H., 1977: J. Appl. Phys. **48**, 2244. Sects. 3.2.1, 5.1, 5.2.1, 5.4, 5.7.2, (Tab. 5.1)

Sørensen, H., J. Schou, Chen Hao-Ming and P. Borgesen, 1983: Surface Sci. **125**, 355. Sects. 5.1, 5.7.2, (Tab. 5.1)

Soszka, W., 1973a: Surface Sci. **36**, 48. Sect. 5.8

Soszka, W., 1973b: Acta Phys. Pol. **A44**, 841. Sects. 5.1, 5.8, (Tab. 5.1)

Soszka, W., 1978: Surface Sci. **74**, 636. Sect. 5.1, (Tab. 5.1)

Soszka, W., 1990: Nucl. Instrum. Meth. Phys. Res. B **48**, 630. Sect. 5.1, (Tab. 5.1)

Soszka, W. and M. Soszka, 1977: Acta Phys. Pol. **A51**, 319. Sects. 5.1, 5.8, (Tab. 5.1)

Soszka, M. and W. Soszka, 1980: Nucl. Instrum. Meth. **168**, 585. Sects. 3.4, 5.1, 5.5, 5.8, (Tab. 5.1)

Soszka, M. and W. Soszka, 1983a: Acta Phys. Pol. **A64**, 255. Sects. 5.1, 5.5, 5.8, (Tab. 5.1)

Soszka, M. and W. Soszka, 1983b: Nucl. Instrum. Meth. **218**, 782. Sects. 5.1, 5.8, 5.10, (Tab. 5.1)

Soszka, M. and W. Soszka, 1983c: Phys. Lett. **97A**, 256. Sects. 5.1, 5.8, (Tab. 5.1)

Soszka, W. and T. Stepien, 1981: Nucl. Instrum. Meth. **182/183**, 163. Sects. 5.1, 5.8, 5.10, (Tab. 5.1)

Soszka, W., A.J. Algra, E.P.Th.M. Suurmyer and A.L. Boers, 1980: Rad. Effects **51**, 171. Sects. 5.1, 5.8, 6, (Tabs. 5.2, 6.1)

Soszka, W., S. Kwasny, J. Budzioch and M. Soszka, 1989: J. Phys.: Condens. Matter **1**, 1353. Sect. 5.1, (Tab. 5.1)

Staudenmaier, G., W.O. Hofer and H. Liebl, 1976: Int. J. Mass Spectrom. Ion Phys. **21**, 103. Sects. 5.1, 5.2.5, 5.3, (Tab. 5.1)

di Stefano, T.H., and D.T. Pierce, 1970: Rev. Sci. Instrum. **41**, 180. Sect. 3.3

Stein, J.D. and F.A. White, 1972: J. Appl. Phys. **43**, 2617. Sects. 5.1, 5.7.2, (Tab. 5.1)

Sternglass, E.J., 1957: Phys. Rev. **108**, 1. Sects. 2, 4.2, 5.2.2, 5.3

Stier, P.M., C.F. Barnett and G.E. Evans, 1954: Phys. Rev. **109**, 973. Sects. 5.2.3

Stolterfoht, N., 1978: in *Topics in Current Physics, Vol. 5*, ed. by I.A. Sellin, (Springer-Verlag, Berlin), p. 155. Sects. 4.3.3, 6

Strong, R. and M.W. Lucas, 1977: Phys. Rev. Lett. **39**, 1349. Sect. 5.6

Suarez, S., G.C. Bernardi, P. Focke and W. Meckbach, 1988: Nucl. Instrum. Meth. Phys. Res. B **33**, 326. Sect. 5.6

Svensson, B. and G. Holmén, 1981: J. Appl. Phys. **52**, 6928. Sects. 3.1, 3.1.2, 5.1, 5.2.1, 5.2.2, 5.4, 5.7.2, 6, (Tabs. 5.3, 6.1)

Svensson, B. and G. Holmén, 1982: Phys. Rev. B **25**, 3056. Sects. 4.3.4, 5.1, 5.2.4, (Tab. 5.3)

Svensson, B., G. Holmén and A. Burén, 1981: Phys. Rev. B **24**, 3749. Sects. 4.3.1, 5.1, 4.3.1, (Tab. 5.3)

Svensson, B., G. Holmén and L. Linnros, 1982: Nucl. Instrum. Meth. **194**, 429. Sects. 5.1, 5.2.2, 5.2.4, (Tab. 5.3)

Tabata, T., R. Ito, Y. Itakawa, N. Itoh and K. Morita: At. Data Nucl. Data Tabl. **28**, 493. Sect. 3.2.1

Taglauer, E., 1990: Appl. Phys. A **51**, 238. Sect. 3.1.1

Tanuma, S., C.J. Powell and D.R. Penn, 1990: J. Vac. Sci. Technol. A **8**, 2213. Sect. 4.1.2

Thomas, E.W., 1984a: Nucl. Fus., Spec. Issue **94**, 94. Sects. 1, 5.3, 6, (Tabs. 1.1, 6.1)

Thomas, E.W., 1984b: Vacuum **34**, 1031. Sect. 4.3.5

Thomas, E.W., 1985: ORNL-6088/V3, Oak Ridge National Laboratory, Oak Ridge, USA , C-1. Sects. 5.2.3, 6, (Tab. 6.1)

Thomson, J.J., 1904: Proc. Cambr. Phil. Soc. **13**, 49. Sect. 1

Thornton, T.A. and J.N. Anno, 1977: J. Appl. Phys. **48**, 1718. Sect. 5.1, (Tab. 5.3)

Thum, F. and W.O. Hofer, 1979: Surface Sci. **90**, 331. Sects. 5.1, 5.2.5, 6, (Tabs. 5.1, 6.1)

Thum, F. and W.O. Hofer, 1980: in *Symposium on Atomic and Surface Physics (SASP)* (Maria Alm, Austria), p. 19. Sect. 5.1, (Tab. 5.1)

Thum, F. and W.O. Hofer, 1984: Nucl. Instrum. Meth.B **2**, 531. Sects. 5.1, 5.3, (Tab. 5.2)

Toburen, L.H., and W.E. Wilson, 1972: Phys. Rev. A **5**, 247. Sect. 4.1.1

Toburen, L.H., W.E. Wilson and H.G. Paretzke, 1982: Phys. Rev. A **25**, 713. Sects. 4.3.5, 6

Tougaard, S. and P. Sigmund, 1982: Phys. Rev. B **25**, 4452. Sects. 4.1.2, 4.3.5

Tung, C.J., J.C. Ashley and R.H. Ritchie, 1979: Surface Sci. **81**, 427. Sect. 4.1.2

Varga, P., 1987: Appl. Phys. A **44**, 31. Sect. 5.9

Veje, E., 1981: Rad. Eff. Lett. **58**, 35. Sects. 5.1, 5.2.4, (Tab. 5.2)

Veje, E., 1982: Nucl. Instrum. Meth. **194**, 433. Sects. 5.1, 5.2.1, 5.2.4, (Tab. 5.2)

Veje, E., 1984: Nucl. Instrum. Meth. B **2**, 536. Sects. 5.1, 5.4, (Tab. 5.2)

Veje, E., 1988: Nucl. Instrum. Meth. Phys. Res. B **33**, 497. Sects. 5.1. 5.8, (Tab. 5.2)

Verhoeven, J., 1978: *Proc. Symposium on Surface Contamination: Genesis, Detection and Control*, (Washington, USA). Sect. 3.1.1

Vernickel, H., 1978: Phys. Rep. **37**, 93. Sect. 6, (Tab. 6.1)

Viel, L., N. Colombie, B. Fagot and C. Fert, 1970: C. R. Acad. Sc. Paris **271 B**, 239. Sect. 5.1, (Tab. 5.2)

Vinokurov, Ya.A., L.M. Kishinevskii and E.S. Parilis, 1976: Bull. Acad. Sci. USSR, Phys.Ser. **40**, 166. Sects. 4.2, 4.3.1

Vriens, L., 1969: in *Case Studies in Atomic Collision Physics I*, ed. by E.W. McDaniel and M.R.C. Mcdowell, (North Holland, Amsterdam), p. 337. Sect. 4.3.6

Wehner, G., 1966: Z. Physik **193**, 439. Sects. 5.1, 5.6, (Tab. 5.1)

Wille, U., and R. Hippler, 1986: Phys. Rep. **132**, 129. Sect. 4.1.1

Wittmaack, K., 1977: in *Inelastic Ion-Surface Collisions*, ed. by N.H. Tolk, J.C. Tully, W. Heiland and C.W. White, (Academic Press, New York), p. 153. Sect. 3.2.1

Wolff, P.A., 1954: Phys. Rev. **95**, 56. Sect. 4.1.2

Wurtz, J.L. and C.M. Trapp, 1972: J. Appl. Phys. **43**, 3318. Sects. 5.1, 5.3, (Tab. 5.2)

Zalm, P.C. and L.J. Beckers, 1984: Philips J. Res. **39**, 61. Sects. 3.1.2, 5.1, 5.2.1, 5.2.4, (Tab. 5.1)

Zalm, P.C. and L.J. Beckers, 1985: Surface Sci. **152/153**, 135. Sects. 5.1, 5.2.1, (Tab. 5.1)

Zampieri, G. and R. Baragiola, 1986: Phys. Rev. B **33**, 588. Sect. 5.6

Zampieri, G., F. Meier and R. Baragiola, 1984: Phys. Rev. A **29**, 116. Sect. 5.6

Ziegler, J.F., J.P. Biersack and U. Littmark , 1985: *The Stopping and Range of Ions in Solids*, (Pergamon Press, New York). Sect. 5.2.2

Kinetic Electron Emission from Ion Penetration of Thin Foils in Relation to the Pre-Equilibrium of Charge Distributions

H. Rothard, K.O. Groeneveld and *J. Kemmler*

With 20 Figures

1. Introduction

The kinetic electron emission from the ion penetration through thin foils and the relation to the pre-equilibrium of the ion charge distribution will be discussed in this review. An ion of energy E_P and of normal incidence onto the surface of a solid looses energy ΔE_P while penetrating a solid foil of thickness d (Fig. 1.1). The foil is called thin in this context if $\Delta E_P < E_P$. Here, foils must be self-supporting and, for practical reasons, these foils cannot be thinner than $d \approx 50$ Å. Again, for practical reasons, this is possible only if the ion has sufficient velocity, i.e., say, $v_P^2 = E_P/M_P > 20\,\mathrm{keV/u}$. Thus, the contribution of potential electron emission decreases strongly. We

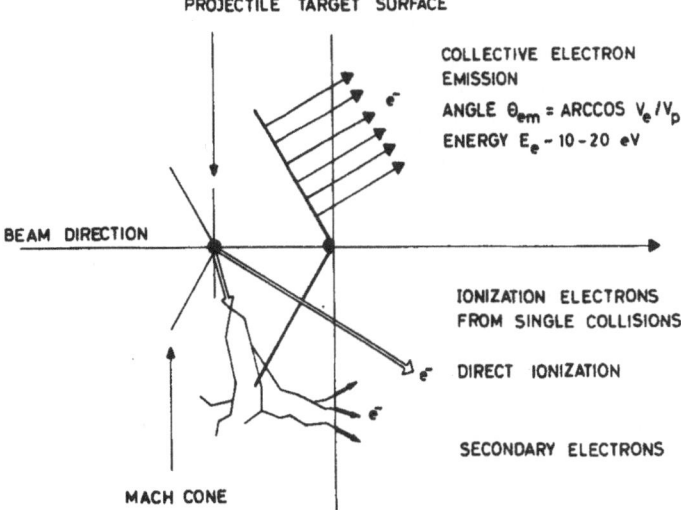

ELECTRONS FROM HEAVY ION-SOLID INTERACTION

Fig. 1.1. Possible mechanisms for electron emission from (heavy) ion interaction with solids as, e.g., i) primary ionization from binary collisions, ii) secondary electrons from cascade multiplication or iii) collective effects like the emission of shock electrons by the ion-induced wake cone (see, e.g., Groeneveld 1988)

restrict, therefore, the discussion in this chapters mostly to the discussion of kinetic electron emission.

The production of such thin foils is in general a rather tedious and complicated procedure for a number of reasons: they must be thin, self-supporting, free of surface contamination, it must be possible to clean the surface and smooth the surface (mostly by sputtering processes), their geometrical, elemental and chemical composition must be known and controlled during the experiment with a high degree of sensitivity and accuracy. Most of the experimental aspects are discussed in the other chapters of this volume. For a detailed discussion of the problems associated with "self-supporting" and "thin" foils and related questions, the reader is referred to Burkhard et al. (1988b) and Lorenzen et al. (1989).

Under such conditions, however, electron emission from thin foils renders unique information. The interaction time between the ion and the foil is in the order of 10^{-18} s to 10^{-14} s, it is well defined and can be controlled accurately, and can be varied over a wide range. Both the electron emission from the entrance and from the exit surfaces become accessible. This, in particular, makes it possible to study with *thin* foils the evolution of pre-equilibrium- to equilibrium-conditions of such processes as electron capture, electron loss, the charge states, the excitation states, the energy loss etc.. The quantities which are accessible experimentally are: energy and angular distributions of electrons under ion impact ($d^2n/dE\,d\Omega$, $N(E)$, $N(\Theta)$, Chaps. 3 and 5), and the number of electrons ejected per incident projectile in the forward hemisphere (γ_F) or the backward hemisphere (γ_B) (Chaps. 2 and 4).

In particular, it becomes possible to relate, via a coincidence technique, the electron emission from the exit surface to one specific charge state q_f of the charge-analysed, emerging ion (Chaps. 4 and 5). A typical experimental setup for the measurement of doubly differential electron spectra $d^2n/dE\,d\Omega$ in coincidence with the final charge state q_f of the ions is shown in Fig. 1.2. The (heavy) ion beam with incoming charge state q_i interacts with the thin target foil. The electron spectra are measured with a magnetic (or an electrostatic) energy analyzer. The different charge states can be separated with an electric or a magnetic field and the charge state distribution can, e. g., be determined by a position sensitive ion detector. For more details see, e. g., Biedermann et al. (1988), Kemmler (1988), Burkhard, Lotz and Groeneveld (1988), Rothard et al. (1991a).

The interaction time $\Delta t = \Delta x/v_P$ of the ion with the solid is in the range of the inverse plasma frequency ω_P^{-1} of the penetrated solid and thus refers to collective excitation modes. Special emphasis will be given to the finger prints of collective excitation in the measured energy and angular distributions of the electrons ejected under ion bombardment (Chap. 3). The electrons born in such *collective excitation* processes (Fig. 1.1, upper half) are contrasted to those born in *binary collisions* (Fig. 1.1, lower half) between the (constituents of the) penetrating particle and target atoms.

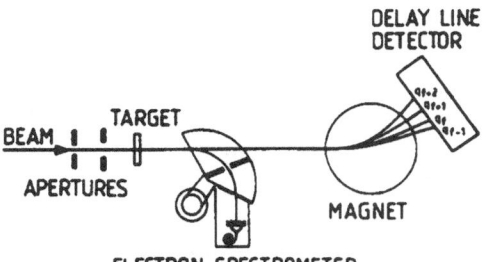

EXPERIMENTAL SET UP

DELAY LINE
DETECTOR

BEAM

TARGET

APERTURES

ELECTRON SPECTROMETER

MAGNET

Fig. 1.2. A typical experimental setup for the measurement of doubly differential electron spectra $d^2n/dE\,d\Omega$ in coincidence with the final charge state q_i of the ions. The (heavy) ion beam with incoming charge state q_i interacts with the thin target foil. The electron spectra are measured with a magnetic (or an electrostatic) energy analyzer. The different charge states can be separated with an electric or a magnetic field and the charge state distribution can, e.g., be determined by a position sensitive ion detector. For more details see, e.g., Biedermann et al. (1988), Kemmler (1988), Burkhard, Lotz and Groeneveld (1988), Rothard et al. (1991a)

The study of these processes may be extended under channeling conditions to selected impact parameters and under grazing incidence, additionally, to a much longer time scale.

2. Forward and Backward Electron Yields in the Charge Equilibrium

The backward "secondary electron" (SE) yield γ_B from swift (i.e. at projectile velocities $v_\mathrm{P} > 0.1v_\mathrm{B}$, v_B being the Bohr velocity) proton and heavy ion bombardment of thick solid targets has been studied as a function of a variety of parameters, such as v_P, the projectile nuclear charge Z_P, the charge state of the ion q_i, the target tilt angle δ (with $\delta = 0°$ denoting perpendicular ion impact), and the target material Z_T (see e.g. the extensive reviews by Hasselkamp (1985, 1988), Brusilovsky (1990), Hofer (1990) and Hasselkamp (1991) and Devooght et al. (1991), this and the preceding volume). Although electron emission from thin foils has first been investigated by Schneider (1931), only a few papers have been published concerning measurements of the total secondary electron yield, $\gamma_\mathrm{T} = \gamma_\mathrm{B} + \gamma_\mathrm{F}$, with thin foil targets (Clerc et al. 1973, Schader et al. 1978, Garnir, Dumont, Baudinet-Robinet 1982, Frischkorn and Groeneveld 1983, Clouvas et al. 1989, 1991). Here γ_F is the yield of SE emitted in the forward hemisphere, i.e. the yield from the surface where the ions exit from the foil. In these experiments, γ_T has been studied as a function of v_P, Z_T, δ and the target thickness d with different projectiles. Also, it has been found that the target temperature may affect γ_T (Gay and Berry 1979, Rothard et al. 1988b).

Some authors have studied the emission of SE separately from both the entrance and the exit surfaces of thin foils and deduced the ratio of forward to backward SE yields, $R = \gamma_F / \gamma_B$. The ratio R, introduced by Meckbach, Braunstein and Arista (1975), has been studied as a function of v_P with different ions and target materials (Pferdekämper and Clerc 1975, 1977, Koyama et al. 1982, Shi et al. 1985, Dehaes, Carmeliet, Dubus 1986, Dednam et al. 1987, Da Silveira and Jeronymo 1987, Kroneberger et al. 1988, Rothard et al. 1989b, Kroneberger et al. 1989, Koschar et al. 1989, Rothard et al. 1990a,b). Although ion-induced electron emission strongly depends on surface properties (compare Sect. 3.3), only two groups have performed experiments with thin foils in ultrahigh vacuum under controlled surface conditions (Meckbach, Braunstein and Arista 1975, Rothard et al. 1989b, Kroneberger et al. 1989, Rothard et al. 1990a, b). Secondary electron emission from molecular ion impact will be discussed in Sects. 3.4 and 4.2.

2.1 Proportionality Between Secondary Electron Yields and Stopping Power

The most important theoretical models consider γ to be proportional to the electronic stopping power S_e,

$$\gamma = \Lambda^* S_e , \tag{2.1}$$

(see, e. g., Sternglass 1957, Schou 1980, 1988, Sigmund and Tougaard 1981, Hasselkamp 1985, Devooght, Dubus, Dehaes 1987, Rösler and Brauer 1988, Koschar et al. 1989, Rothard et al. 1990a, 1990b). Thus, it became common practice to define parameters Λ^* as ratios between the measured SE yields and the (tabulated) stopping power values to study the validity of this assumption as a function of v_P, Z_P and Z_T:

$$
\begin{aligned}
\Lambda_T^* &= \gamma_T / S_e \quad \text{for the } \textit{total} \text{ SE yield} \\
\Lambda_B^* &= \gamma_B / S_e \quad \text{for the } \textit{backward} \text{ SE yield} \\
\Lambda_F^* &= \gamma_F / S_e \quad \text{for the } \textit{forward} \text{ SE yield} .
\end{aligned}
\tag{2.2}
$$

For proton bombardment, however, (2.1) has been confirmed experimentally for a number of target materials within an accuracy of about $\Delta\Lambda^* < \pm 8\%$ in a wide projectile energy range, i.e. $10\,\text{keV} \leq E_P \leq 10\,\text{MeV}$ for γ_B (Au, Ag, Cu, Al) (Hasselkamp 1985, 1988) and $5\,\text{MeV} \leq E_P / M_P \leq 24\,\text{MeV}$ for γ_B (Al_2O_3, Au) (Borovsky, McComas, Barraclough 1988) as well as $400\,\text{keV} \leq E_P \leq 10\,\text{MeV}$ for γ_T (C) (Clouvas et al. 1989). In this case, i.e. for protons, the parameters Λ_T^* and Λ_B^* could then be expected to depend only on target properties such as ionization cross sections σ, transport lengths λ for secondary particles in condensed matter, and the surface-dependent escape probabilities. This would mean that $\Lambda_{T,B}^*$ may be considered as "*material parameters*" for the particular case of proton impact. For heavier targets (Ti, Ni, Cu) and lower proton energies ($E_P \leq 300\,\text{keV}$), Λ_F^*

is enhanced compared to the value of Λ^* at $E_P > 300\,\text{keV}$, but this is not the case for the light targets C and Al (Rothard et al. 1989b).

With heavy ions, deviations from the simple rule (2.1) have been observed, especially at low projectile velocities, i.e. $v_P^2/2 = E_P/M_P < 150\,\text{keV/u}$ (Holmen et al. 1979, Frischkorn and Groeneveld 1983, Hasselkamp 1985, Rothard et al. 1989b, 1990a). Λ_B^* and Λ_F^* were found to depend on the projectile atomic number Z_P (Holmen et al. 1979, Rothard et al. 1990, Clouvas et al. 1991 and Sect. 2.3) and, in accordance with the establishment of charge equilibrium and the consequent adjustment of stopping powers close to the entrance surface of the solid target, on the charge state of the incoming projectiles, or the "partial stopping powers" that are characteristic for these charge states (see Koschar et al. 1989 and Chap. 4). Also, it is an open question whether the proportionality (2.1) holds for molecular ion impact (see, e. g., Hasselkamp 1985, Kroneberger et al. 1989, Rothard et al. 1990b and Sects. 3.4 and 4.2).

As an example for the v_P- and Z_P-dependence of SE yields and the Λ parameters, the lower part of Fig. 2.1 shows the SE yields γ_B (left) and γ_F (right) for a copper-target as a function of the projectile energy per unit mass $E_P/M_P = v_P^2/2$ ($15\,\text{keV/u} \leq E_P/M_P \leq 600\,\text{keV/u}$). The charge state of the incoming ions was $q_i = 1$ (Rothard et al. 1990a). The upper parts of Fig. 2.1 show the $E_P/M_P = v_P^2/2$-dependence of the parameters

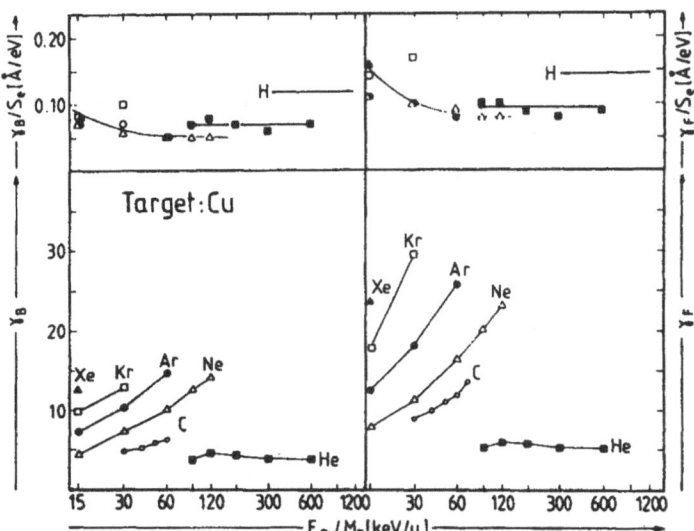

Fig. 2.1. Heavy ion-induced (Xe, Kr, Ar, Ne, C, He, $q_i = 1$) secondary electron yields (lower part) from the sputter-cleaned entrance (γ_B, *left*) and exit surfaces (γ_F, *right*) of a thin Cu foil ($d \approx 1000\,\text{Å}$) as a function of the square of the projectile velocity $1/2v_P^2 = E_P/M_P$ (from Rothard et al. 1990a). The upper parts show the corresponding secondary electron yield to projectile energy loss ratios $\Lambda^* = \gamma/S_e$ (full triangles: Xe, open squares: Kr, full circles: Ar, open triangles: Ne, full squares: He). The solid lines are to guide the eye and represent mean values in the case of Λ^* (He). Also, mean Λ^* values for H projectiles (300 to 1200 keV/u) have been included in the figure

$\Lambda_{F,B}^* = \gamma_{F,B}/S_e$. The stopping power values have been calculated according to Ziegler, Biersack and Littmark (1985) (ZBL). $\Lambda_{F,B}^* (v_P)$ are independent of v_P at $E_P/M_P > 50\,\text{keV/u}$, i.e. with respect to the dependence on v_P, both forward and backward SE yields follow the velocity-dependence of the projectile energy loss even in the present case of heavy ion impact.

At lower velocities, γ is higher than expected from (2.1) if tabulated (ZBL) electronic stopping power values are used. One important reason for the enhancement may be the *contribution of recoil ion cascade induced SE* to the total SE yield. This phenomenon is related to the nuclear stopping and thus should become more important with decreasing v_P. The contribution of recoils has been investigated in detail experimentally and has been described with a semiempirical model by Holmen et al. (1979). Other possible explanations have been mentioned by Frischkorn and Groeneveld (1983), e.g. a *transition from the cascade electron regime to single ionization processes* (see also the discussion for the v_P-dependence of R, Sect. 2.2), or *enhanced molecular orbital ionization in heavy symmetric collision systems*. This would alter the internal SE energy distribution, in contrast to the assumptions of the theoretical models (compare Sect. 2.3), where the shape of the SE energy distribution comes out to be independent of the projectile velocity or charge, which both only determine the absolute value as scaling factors. Further conclusions can only be drawn from measurements of the energy and angular distributions of heavy ion-induced SE both at low and high projectile velocities (keV/u to some $10\,\text{MeV/u}$ range, compare Sect. 2.1).

There is a remarkable dependence of γ_F and γ_B on the projectile atomic number at a given velocity and for a given target material: Both SE yields increase strongly with Z_P. Also, in contrast to (2.1), $\Lambda_{F,B}^*$ depend on the projectile nuclear charge Z_P. The "saturation values" of $\Lambda_{F,B}$ are systematically lower for the heavier ions compared to He and H, i.e.

$$\Lambda^*(\text{Ne}) < \Lambda^*(\text{He}) < \Lambda^*(\text{H}) \tag{2.3}$$

(Rothard et al. 1989b, 1990a) for γ_F, γ_B and γ_T and different target materials (C, Al, Ti, Ni, Cu)! A possible understanding of these results in terms of an effective near-surface energy loss is discussed in Sect. 2.3.

Finally, if Λ is expressed in units of Å/eV rather than in units of $(\mu\text{g/cm}^2)/\text{eV}$, i.e. by taking into account the different densities of the targets and relating γ to the "energy loss per unit path length", dE/dx, and not to the "stopping cross section", $(dE/dx)/N$, very similar Λ values for all the studied target materials are found. Rothard et al. (1990a) deduced a mean value of $\Lambda_T = \Lambda = 0.32\,\text{Å/eV}$ with $Z_T = 6, 22, 28, 29$ in good agreement with mean values of Λ given by Clouvas et al. (1989) ($\Lambda = 0.31\,\text{Å/eV}$; $Z_T = 28, 29, 46, 47, 61, 63, 79, 83$) and Schou (1988) ($\Lambda = 0.29\,\text{Å/eV}$; $Z_T = 4, 12, 13$).

In spite of the systematic deviations concerning the Z_P- and Z_T-dependences discussed in Sects. 2.2 and 2.3, the important assumption of an

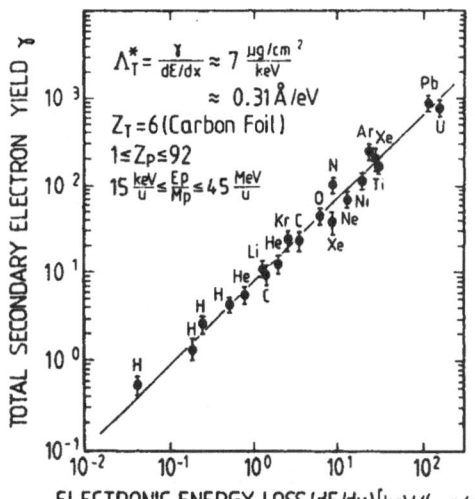

Fig. 2.2. The total secondary electron yield γ_T from carbon foils as a function of the electronic energy loss dE/dx of the projectiles. The projectiles (15 keV/u $\leq E_P/M_P \leq$ 45 MeV/u, $1 \leq Z_P \leq 92$) are indicated in the figure

overall proportionality between SE yields and the electronic energy loss of the projectiles (2.1) is demonstrated impressively in Fig. 2.2, which shows the total secondary electron yield γ_T from carbon foils as a function of the electronic energy loss. Similar plots have been given by Frischkorn and Groeneveld (1983) and Rothard et al. (1990a), here enriched with further data. A rough proportionality $\gamma_T \sim dE/dx$ within a factor of 2 in a wide range of projectile velocities v_P (15 keV/u $\leq E_P/M_P \leq$ 46 MeV/u) and projectile nuclear charges Z_P ($1 \leq Z_P \leq 92$) over four decades of secondary electron yields γ and electronic energy loss values dE/dx can be stated. Again, the mean value of the proportionality factor is $\Lambda_T^* = 0.31$ Å/eV. The deviations within a factor 2 from this mean material parameter can be attributed to the Z_P- and Z_T-dependence of Λ^*, whereas the velocity-dependence $\Lambda^*(V_P) = $ const. is confirmed, see above.

According to Rothard et al. (1990a), these findings allow a simple relationship for an estimate of secondary electron yields γ_B and γ_F from metallic solids and solid foils for H, He and Ne and Ar ion (HI) impact:

$$\gamma_B = 0.14 C_B \, dE/dx \,, \quad \text{with } C_B(\text{H}) = 1\,, \ C_B(\text{He}) = 0.6\,, \ C_B(\text{HI}) = 0.3$$
$$\gamma_F = 0.17 C_F \, dE/dx \,, \quad \text{with } C_F(\text{H}) = 1\,, \ C_F(\text{He}) = 0.65\,, \ C_F(\text{HI}) = 0.5$$
$$(2.4)$$

with dE/dx measured in units of eV/Å. Equation (2.4) can be applied in the specific energy range above, say, 50 keV/u and can be expected to give reasonable results up to the MeV/u range, and should be accurate within a factor of 2 for ions with $Z_P \leq 36$.

2.2 Forward to Backward Secondary Electron Yield Ratio

In Fig. 2.1, the ratio $R = \gamma_F/\gamma_B$ is plotted as a function of E_P/M_P for a copper target. In agreement with all previous studies of SE emission following ion penetration of thin foils for a given projectile velocity, γ_F is always higher than γ_B for all projectile/target combinations, i.e. the SE emission in forward direction dominates ($R > 1$). The enhancement of $\gamma_F = 1.2\,\gamma_B$ for protons and a corresponding fraction of the enhancement for heavier ions is caused by additional SE creation in forward direction by fast δ-electrons. It has been observed generally that R either increases with v_P until a constant value is reached at velocities corresponding to $E_P/M_P \approx 100\text{–}200\,\mathrm{keV/u}$ or that it remains constant also at lower projectile velocities (Meckbach, Braunstein, Arista 1975, Shi et al. 1985, Dednam et al. 1987, Rothard et al. 1990a).

From Fig. 2.3, the important conclusion that the ratio $R = \gamma_F/\gamma_B$ increases with increasing projectile nuclear charge Z_P by up to a factor of two in the range $1 \leq Z_P \leq 54$ can be drawn. This important finding is demonstrated impressively in this figure, where mean values of R (Al, Ni, Cu) are plotted as a function of Z_P ($Z_P = 1, 2, 10, 18, 36, 54$). According to Rothard et al. (1990a) and following Sternglass (1957) and Da Silveira and Jeronymo (1987), the forward/backward emission ratio $R = \gamma_F/\gamma_B$ can be expressed as

$$R = \frac{\gamma_F}{\gamma_B} = \frac{S_{eF}^*}{S_{eB}^*} \left[1 - \frac{B}{1 + (\lambda_{SE}/\lambda_\delta)} \right]^{-1} . \tag{2.5}$$

The quantities S_e^* are the effective near-surface projectile energy losses (Sect. 2.3) and are proportional to the square of the effective charges $(q_{eff})^2$

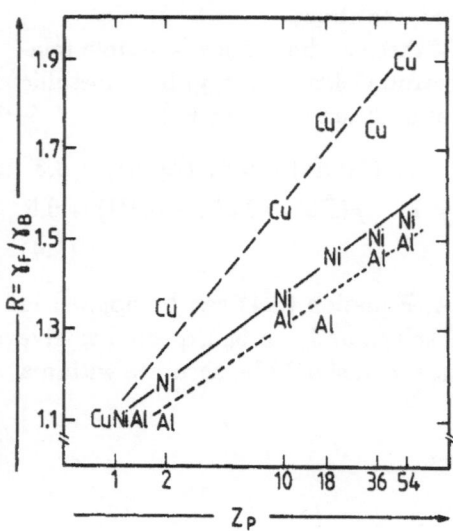

Fig. 2.3. The mean values of the forward and backward secondary electron yield ratio $R = \gamma_F/\gamma_B$ as a function of the nuclear charge Z_P for different target materials (Al, Ni, Cu) as indicated in the figure (from Rothard et al. 1990a). The lines are to guide the eye

of the ions near the surface. The factor B describes the partition of the projectile energy loss in two different types of collision processes,

1) *close collisions* between the projectile and the electrons leading to – mostly forward directed – high-energy δ-electrons (fraction BS_e) and

2) *distant collisions* corresponding also to long-range collective excitations leading to – mostly isotropic – low-energy SE (fraction $(1-B)\,S_e$).

Referring to Bohr (1948), an equipartition $B = (1-B) = 0.5$ was chosen in the original Sternglass (1957)-Ansatz. The corresponding attenuation length λ_δ and λ_{SE} are characteristic for the high-energy δ-electrons or the low-energy SE, respectively. Typical values are $\lambda_{S_E} \approx 15\,\text{Å}$, $\lambda_\delta \approx 300\,\text{Å}$ and $B \approx 0.6$ for C projectiles at a specific projectile energy of $1\,\text{MeV/u}$ (Koschar et al. 1989) and $B(\text{He}) \approx 0.45$ and $B(\text{Ne}) \approx 0.42$ at $E_P/M_P \approx 0.1\,\text{MeV/u}$, i.e. values close to the equipartition $B = 0.5$.

Equation (2.5) furnishes at least a possibility for a qualitative understanding of the velocity-dependence of R. According to (2.5), R increases with increasing v_P and reaches a saturation value at high velocities with $\lambda_\delta \gg \delta_{SE}$. According to Sternglass, the ratio $\lambda_\delta/\lambda_{SE}$ can be estimated as

$$\lambda_\delta/\lambda_{SE} \approx 5.4\,E_P/M_P \qquad (2.6)$$

(with E_P/M_P measured in units of MeV/u). Thus, from (2.5), a saturation of $R(v_P)$ can be expected if $\lambda_\delta > \lambda_{SE}$, i.e. for velocities $1/2v_P^2 = E_P/M_P > 200\,\text{keV/u}$. This seems to be a reasonable agreement with the above mentioned experimental findings indicating a saturation of $R(v_P)$ at projectile velocities around $E_P/M_P \approx 100$ to $200\,\text{keV/u}$ considering that the Sternglass (1957)-theory is strictly valid at high projectile velocities $v_P > Z_P^{2/3} v_B$ (see the discussion in Rothard et al. 1990a). The weak Z_T-dependence of R may have an explanation in a Z_T-dependence of the electron transport lengths λ_{SE} and λ_δ. The important conclusion is that *SE transport in solids* (Chap. 5) is of major importance for the quantitative description of the projectile velocity dependence of SE yields.

R (and also γ_F and γ_B) show a weak dependence on Z_T, but this is not a simple function of Z_T as, e.g., an increase or decrease with Z_T. The dependence of both γ_B and γ_T on the target atomic number Z_T (for a given projectile velocity) has been studied with protons, molecular ions and heavy ions. An oscillatory behavior of $\gamma(Z_T)$ has been observed both for thick samples (γ_B, Hasselkamp 1985, Hippler 1988, Hippler, Hasselkamp, Scharmann 1988, Hasselkamp 1991) and thin foils (Clouvas et al. 1989). Differences in the "fine structure" of these oscillations found for γ_B compared to γ_F must be attributed to a difference in the additional contribution to SE emission from δ-electrons in forward direction. This finding may be caused by a Z_T-dependence of all the quantities having an influence on SE emission, such as, e.g., dE/dx related to the cross sections for *ionization processes*, *electron transport mean free paths* λ, and the *surface potential barrier*. Furthermore, *shell and band structure effects* (Thum and Hofer

1984, Hasselkamp 1985, Hippler, Hasselkamp, Scharmann 1988, Koschar et al. 1989, Hofer 1990), as well as *collective excitations* like plasmon decay (Hasselkamp 1985, Burkhard, Rothard, Groeneveld 1988, Hippler 1988) and "Wake"-effects (Burkhard et al. 1987a, Kroneberger et al. 1989, Rothard et al. 1990b) may have to be considered. In particular, the different valence electron configuration of the different target materials may play an important role for both energy loss and SE emission. All this also holds for the Z_T-dependence of R.

2.3 Projectile Dependence of Electron Yields and the Relation to the Effective Near-Surface Energy Loss for the Projectiles

In the following, it will be shown that the Z_P-dependence of Λ^* and R can be described within the framework of Schou's theory for SE emission (Schou 1980, Schou 1988). According to this transport theory, the (fast) proton-induced SE yields are given by

$$
\begin{aligned}
\gamma_F &= \Lambda \beta_F S_e \\
\gamma_B &= \Lambda \beta_B S_e \\
\gamma_T &= \Lambda S_e \, .
\end{aligned}
\tag{2.7}
$$

The difference in the forward and backward emission is due to energy transport by recoiling electrons away from the entrance surface or into the region near the exit surface described by the dimensionless factors

$$
1 - \beta_F = \beta_B = \beta \, .
\tag{2.8}
$$

From (2.7) one obtains for fast proton (!) impact

$$
R = \gamma_F / \gamma_B = \beta_F / \beta_B = (1 - \beta) / \beta \, .
\tag{2.9}
$$

Thus, one can easily obtain β by measuring the forward to backward SE yield ratio R for fast protons where $q_F^* = q_B^* = 1$. From (2.7–9) follows

$$
\Lambda = \Lambda_F^*(\mathrm{H}) / (1 - \beta) = \Lambda_B^*(\mathrm{H}) / \beta = \Lambda_T^*(\mathrm{H}) \, .
\tag{2.10}
$$

The proton data shown in Fig. 2.3 yield $R \approx 1.2$ and $\beta = \beta_B \approx 0.45$ and $\beta_F \approx 0.55$ (2.10). According to (2.10), the *material parameter* Λ can be obtained by measuring proton-induced total SE yields $\gamma_T = \gamma_B + \gamma_F$ from thin foils in a *more fundamental and convenient way* than by measuring γ_B from thick samples. In the latter case, uncertainties may arise because a detailed knowledge of the factor β is needed to calculate the material parameter Λ. Material parameters for a variety of thin foil targets with sputter-cleaned surfaces have recently been published by Clouvas et al. (1989).

As it is of common interest to use such an universal material parameter Λ, the transport theory can easily be extended to $Z_P > 1$ with $\Lambda(Z_P) =$

const. by introducing Z_P-dependent factors $C_F(Z_P)$ and $C_B(Z_P)$ in (2.7) (Rothard et al. 1990a):

$$\gamma_F = \Lambda(Z_T)(1 - \beta)C_F(Z_P)(S_e)$$
$$\gamma_B = \Lambda(Z_T)\beta C_B(Z_P)S_e . \tag{2.11}$$

The factors

$$C_F = \Lambda_F^*(Z_P \geq 1)/\Lambda_F^*(Z_P = 1)$$
$$C_B = \Lambda_B^*(Z_P \geq 1)/\Lambda_B^*(Z_P = 1) , \tag{2.12}$$

describe the difference between the *tabulated bulk energy loss* values (ZBL) and a "non-equilibrium near-surface stopping power" both at the entrance and the exit surface of the foils. The absolute values have already been given in (2.4). The parameters C describe in a very general way a variety of physical mechanisms that can possibly cause a projectile dependence of an effective energy loss near the entrance or exit surface as, e. g. *charge exchange, screening effects, projectile excitation or ionization* or even *molecular orbital excitation*.

However, the *pre-equilibrium -evolution of the* effective *projectile charge* q^* near the entrance surface, but also at the exit surface, where a *sudden change of the projectile screening* or a *de-excitation of the projectile* having been *excited inside the solid* can take place, are the key for an understanding of the dependence of the near-surface stopping power.

$$S_B^* = C_B(Z_P)S_{ZBL}$$
$$S_F^* = C_F(Z_P)S_{ZBL} , \tag{2.13}$$

on Z_P. The SE yield is determined by the effective ion charges very close to the surfaces within a depth comparable to low energy electron escape depths λ_{SE},

$$\gamma \sim S_e(\lambda_{SE})^* \sim q^*(\lambda_{SE})^2 . \tag{2.14}$$

According to Rothard, Schou and Groeneveld (1991), one can compare the effective ion charges of light ions (H, He) and heavier ions (N) at both the backward (q_B^*) and the forward (q_F^*) side of the foils to the ZBL (1985) values (q_{ZBL}) for carbon targets and projectile velocities of $v_P \approx 0.2$ MeV/u in the following way: The forward effective charge values q_F^* can be calculated from charge state distributions (Biedermann et al. 1988, Žaikov et al. 1986) under the assumption that the effective charge determining the stopping power at the forward surface is not very different from the mean charge of the ions. The backward effective charges q_B^* were calculated by assuming that the escape depth of low energy SE is about $\lambda \approx 15$ Å (Koschar et al. 1989). From the target-thickness dependence of the mean charge $\bar{q}$ of N^{2+} ions (Žaikov et al. 1986) one can – as a first approach – assume that $q_B^* = \bar{q}(d = 15$ Å$)$. However, because the initial charge state of the N ions was $q_i = +2$ and the charge state of the ions used in the experiments under investigation was $q_i = +1$, this procedure may lead to an overestimation of q_B^* and thus to an underestimation of possible deviations of S^* to S_{ZBL}. In the case of H^+ and He^+, the charge state fractions have reached an equilibrium around $d \approx 50$ Å. To obtain a first approximation for q_B^*, a linear approximation of $\bar{q}(d)$ seems to be reasonable.

The energy loss ratios $r_B = S_B^*/S_{ZBL}$ and $r_F = S_F^*/S_{ZBL}$, i.e., according to (2.14) the ratios of the squares of the effective near-surface charges q_F^* and q_B^* to the squares of q_{ZBL} are shown in Table 2.1. These quantities are a direct measure of the deviation of the near-surface energy loss S^* from the calculated ZBL energy loss S_{ZBL}. A comparison to the C_F and C_B values from (2.4) shows that the tendency of the Z_P-dependence of the C values as well as their absolute magnitude are represented within $\pm 40\%$. By taking into account that a fraction of the ion energy loss may lead to either target or projectile ionization, the Z_P-dependence of the C factors can be estimated even more exactly (Rothard, Schou, Groeneveld 1991). Thus, the Z_P-dependence of $\Lambda_F^*(Z_P) = \gamma_F/S_{ZBL}(Z_P)$ and $\Lambda_B^*(Z_P) = \gamma_B/S_{ZBL}(Z_P)$ can be well understood by assuming an effective charge q^* which determines the non-equilibrium stopping power S^* of the ions near the surface. It has often been proposed to perform simultaneous measurements of γ and S_e. The discussion above shows, however, that only simultaneous measurements of γ and S^* may be meaningful.

The dependence of q^* on Z_P results in a depression of SE yields with increasing Z_P. Interestingly, such a suppression of low energy electrons under heavy ion impact can be seen both for SE from primary ionization (single collision conditions, see, e. g. Schader et al. 1987, Toburen 1991) and for (cascade?) electrons from solids (Koyama et al. 1988, Folkmann et al. 1975). A variety of different mechanisms have been discussed in this context, but it is still unclear in which magnitude each of them contributes to the low energy electron suppression.

The most important mechanisms are probably the *screening of the projectile nuclear charge by projectile electrons* (Toburen 1991, even projectile shell effects have been observed, compare Koschar et al. 1989) and the *screening of the projectile charge by target electrons* in metals (Koyama et al. 1988). Also, changes of the surface barrier height caused by a *charging-up near the ion track* (Koyama et al. 1982), an *interaction of the ions' wake*

Table 2.1. Absolute values of the C and r factors (see text)

Projectile	Z_P	C_B	r_B	C_F	r_F
H	1	1.00	1.00	1.00	1.00
He	2	0.60	0.45	0.65	0.82
N	7		0.47		0.67
Ne	10	0.32		0.49	

with the surface potential (Frischkorn et al. 1982), or a *depression of the SE excitation probability due to a high density of electron-hole pairs* (which then do not longer remain uncorrelated) have been mentioned (Koyama et al. 1988), compare also Rothard et al. (1990a) for a more detailed discussion.

The concept of an *effective near-surface energy loss* is closely related to a novel approach to describe non-equilibrium SE yields within an extended Sternglass (1957)-type model (Koschar et al. 1989) (Chap. 4). This concept is strongly supported by the charge-state dependence of SE yields (Koschar et al. 1989) and recent calculations of the dependence of the energy loss of ions on the charge state of the ions inside the solid (Arnau et al. 1990).

3. Electron Energy and Angular Distributions in the Charge Equilibrium

3.1 Electron Spectra as a Function of the Emission Angle

A typical SE velocity spectrum $d^2n/dE\,d\Omega$ (observation angle $\Theta = 0°$) is shown in Fig. 3.1 for protons (1.2 MeV) traversing a sputter-cleaned carbon foil. The following structures can be identified (Groeneveld, Maier and Rothard 1990):

1) The dominating structure is the low energy "true" SE peak with an intensity maximum at (2.1 ± 0.3) eV. This peak contains about 85 % of all SE and thus the electron yields γ discussed in the previous chapter mainly mirror the behavior of these low energy electrons.

2) The prominent convoy electron (CE) peak appears at an electron vector velocity equal to the projectile vector velocity, $v_e = v_P$ (Groeneveld, Meckbach and Sellin 1984). The strong dependence of the shape of both the

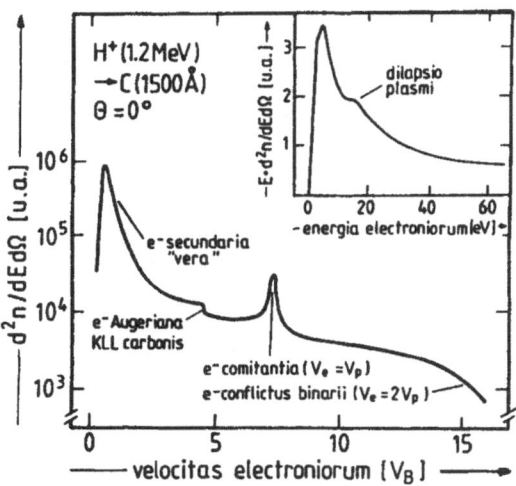

Fig. 3.1. Typical secondary electron spectrum ($\Theta = 0°$) from a carbon foil penetrated by protons (1.2 MeV) according to Groeneveld, Maier, Rothard (1990). The structures indicated in the figure as "true" SE ("e⁻ secundaria vera"), Auger electrons ("e⁻ Augeriana"), convoy electrons ("e⁻ comitantia"), binary encounter electrons ("e⁻ conflictus binarii") or electrons from plasmon decay ("dilapso plasmi") are discussed in the text

low energy "true" SE energy distribution and the CE peak on the coverage of the surfaces with adsorbates is discussed in detail in Sect. 3.3.

3) At $E_e \approx 270\,\text{eV}$, the carbon KLL Auger electron distribution is found. It may serve as an important tool in surface analysis (Burkhard et al. 1988a, Lorenzen et al. 1989).

4) At twice the projectile velocity, $v_e \approx 2v_P$, the high-energy binary encounter electron distribution from close collisions between the projectile and a target electron can be seen. The energy and angular distribution of such high-energy δ-electrons in collisions of Ne^{q+} ($q = 7, 10$, $E_P = 70$–$170\,\text{MeV}$) has recently been investigated both experimentally and theoretically (Schiewitz et al. 1990). For sufficiently high electron energies ($v_e > v_P$), calculated doubly differential electron distributions $d^2n/dE\,d\Omega$ agree well with the experimental data. This model describes i) the production of SE by using atomic ionization cross sections and ii) the transport by assuming a separation of energy loss and angular scattering (compare also Chap. 5).

5) At the top of Fig. 3.1 the low energy part of the SE spectrum $Ed^2n/dE\,d\Omega$ is shown. At an energy of $E_e < 20\,\text{eV}$ a structure is observed which is attributed to the decay of a collective excitation of the electron plasma of a solid, the plasmon. In this process, the energy $\hbar\omega_P$ of a plasmon (for carbon: $\hbar\omega_P \approx 25\,\text{eV}$) is transferred to a single electron. When escaping from the surface, the energy of these electrons is reduced by the surface potential given by the work-function Φ (for carbon: $\Phi \approx 5\,\text{eV}$). Eventually, they are observed at an energy of

$$E_e < \hbar\omega_P - \Phi\,, \tag{3.1}$$

i. e. at $E_e < 20\,\text{eV}$ in the case of carbon. Recently, the decay of plasmons from heavy ion impact on C, Al and Cu foils has been reported (Burkhard, Rothard and Groeneveld 1988). Similar structures in ion- induced electron spectra from thick samples that have been attributed to the decay of plasmons have been observed for Al and Mg (Hasselkamp, Hippler and Scharmann 1987) and Mg, Si, Ti, Ni and Au (Hippler 1988, Hasselkamp 1991). Electrons from the decay of plasmons can only be observed from clean surfaces (Hasselkamp 1985, Burkhard, Rothard and Groeneveld 1988, Rothard et al. 1991a).

Figure 3.2 presents SE velocity spectra $d^2n/dE\,d\Omega$ taken at three observation angles ($\Theta = 0°$, $\Theta = 65°$ and $\Theta = 180°$) from sputter-cleaned gold foils bombarded with H^+ and H_2^+ (0.8 MeV/u, normalized to the same number of protons). The $0°$-spectra show similar structures 1–5 as the one shown in Fig. 3.1 (see also Oda et al. 1980, Yamazaki 1988, Suarez et al. 1988): the dominating "true" SE (A), the CE (C) and the binary encounter electrons (D). A closer investigation of the spectra shows that gold Auger electrons (B) can be observed at energies $E_e \approx 40, 70, 140, 270\,\text{eV}$ resulting from various transitions (e. g. OVV, NVV, NNV). The $65°$ spectra do not, of course, show the convoy electron peak. The intensity of the forward

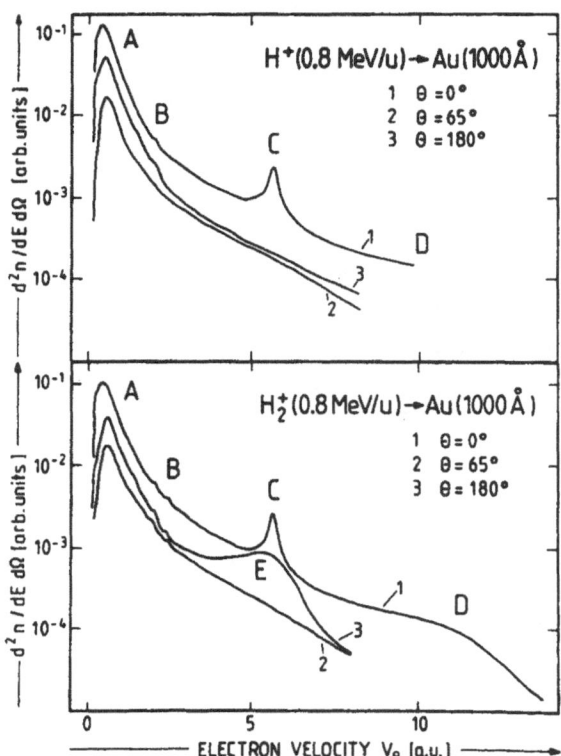

Fig. 3.2. Secondary electron spectra from H⁺ (*top*) and H₂⁺ (*bottom*) bombardment of sputter-cleaned gold foils (Rothard et al. 1990c). The spectra have been recorded at different observation angles: $\Theta = 0°$ labeled 1, $\Theta = 65°$ labeled 2 and $\Theta = 180°$ labeled 3. The structures labeled A to E are discussed in the text

(i.e. $\Theta = 0°$) spectra is higher by a factor of about 2 than the intensity of the backward SE spectra (i.e. $\Theta = 180°$): the intensity of the 65° forward-spectra is much lower than the intensity of the 0°-spectra.

6) The 180°-spectra from H₂⁺ molecular ion impact show a broad structure (E) around $v_e \approx -v_P$ (v_P corresponds to an equivalent electron energy of $E_{eq} = v_P^2 m_e / 2 = 435\,\mathrm{eV}$). This peak originates from collisional loss of projectile electrons, see Sect. 3.4.

In Fig. 3.3 doubly differential secondary electron distributions $N(E) = d^2n/dE\,d\Omega$ from H⁺ ($v_P = 4.0v_B$) and C⁺ ($v_P = 1.2v_B$ and $v_P = 2.7v_B$) bombardment of Au foils (thickness $d = 1000\,\text{Å}$) are presented. The spectra are normalized to $N(E_{max}) = 1$ (with E_{max} being the energy of the maximum of $N(E)$). The left-hand side of Fig. 3.3 shows a comparison of spectra obtained with C⁺ (top) and H⁺ (center) at an observation angle of $\Theta = 120°$ (backward direction) with perpendicular incidence of the beam (target tilt angle $\delta = 0°$). The bottom of Fig. 3.3 shows the ratio of the above two spectra. Qualitatively, the spectra are very similar, i.e. the low energy maximum E_{max} is found at similar energies and also the FWHM is quite similar. These findings are in good agreement with previous results of angle-integrated SE spectra from thick targets in backward direction (Hasselkamp, Hippler, Scharmann 1987), as can be seen from the comparison presented in Table 3.1.

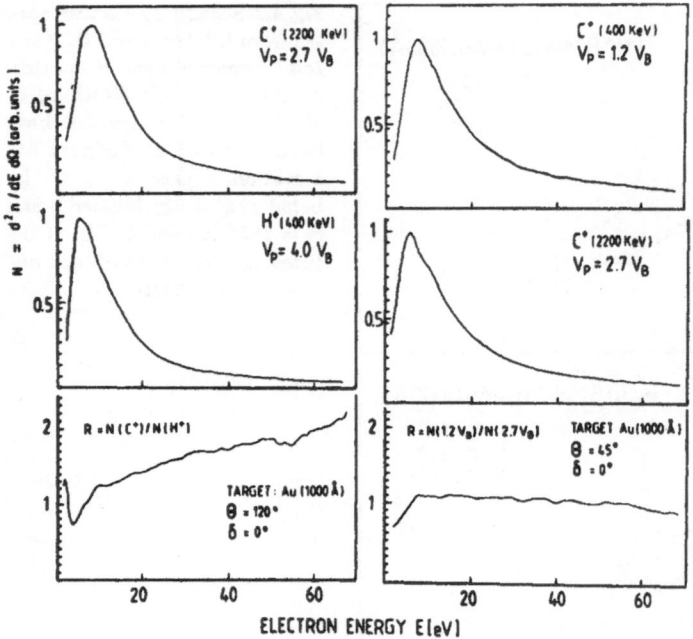

Fig. 3.3. Electron spectra from Au foils bombarded with H^+ and C^+ (the projectile velocities and energies as well as the observation angle Θ are indicated in the figure) (*middle* and *top*). The bottom part shows the ratio R of the above two spectra

The right-hand side of Fig. 3.3 presents spectra obtained with C^+ at two velocities $v_P = 1.2 v_B$ and $v_P = 2.7 v_B$ in forward direction ($\Theta = 45°$). Within experimental uncertainties, the shape of the spectra is independent of v_P above $E > 4\,\text{eV}$. This can clearly be seen from the ratio of the spectra shown in the bottom of Fig. 3.3. Thus, there is no significant velocity dependence of the shape of the spectra. Also, the spectra are quite similar to those observed in backward direction (left). The absolute magnitude of the spectra should however scale according to the energy loss of the ions.

At this point, it is important to note that the detection efficiency of the spectrometer rapidly decreases at low electron energies ($E < 10\,\text{eV}$). Thus, the values for E_{max} and FWHM differ significantly from the values obtained with different experimental procedures (compare, e. g., Hasselkamp, Hippler, Scharmann 1987). However, the ratio of spectra of different ions obtained under the same experimental conditions is independent of the detection efficiency of the spectrometer and thus yields reliable and interpretable information.

The quantities characterizing the shape of the SE energy distribution (E_{max}, FWHM) for e^-, H^+ and heavy ion impact on Au are summarized in Table 3.1. The data for angle-integrated spectra are taken from Hasselkamp, Hippler, Scharmann (1987) (H^+ and Ar^+) and Bindi (1978) (e^-). Although all the energy distributions look alike, there is a slight ten-

Table 3.1. Comparison of the quantities E_{max} and FWHM characterising the shape of particle-induced SE energy distributions. Emission-angle integrated data are denoted by "int." and are taken from Hasselkamp, Hippler, Scharmann (1987) (HHS) and Fig. III.8 of Bindi (1978).

Projectile	E_P [keV]	v_P [v_B]	Θ[deg]	E_{max} [eV]	FWHM [eV]	Reference
e^-	0.6	6.7	int.	1.85	5.5	Bindi 1978
H^+	400	4.0	120	6.0	11.5	Fig. 3.3
	400	4.0	int.	2.2	8.0	HHS 1987
	800	5.7	int.	2.2	8.0	HHS 1987
C^+	400	1.2	45	6.5	14.5	Fig. 3.3
	2200	2.7	45	6.0	14.5	Fig. 3.3
	2200	2.7	120	7.5	12.5	Fig. 3.3
Ar^+	100	0.4	int.	3.0	8.0	HHS 1987
	900	1.0	int.	3.0	12.0	HHS 1987

dency that $E_{max}(e^-) < E_{max}(H^+) < E_{max}$(Heavier ions) and FWHM($e^-$) $<$ FWHM(H^+) $<$ FWHM(Heavier ions). There is no significant velocity dependence for all the different projectiles at sufficiently high velocities ($v_P > v_B$). For a detailed comparison of e^- and H^+ induced spectra see Schou (1988) and Musket (1975).

It is interesting to note from Fig. 3.3 a slightly increasing ratio of $R(E) = N(C^+)/N(H^+)$ with increasing E. This is in good agreement with previous findings, where SE spectra from heavy ion impact were compared to spectra from light ion impact (Folkmann et al. 1975, Koyama et al. 1988). Possibly, these findings may be explained by screening of the projectile charge by projectile- (Folkmann et al. 1975) or even target-electrons (Koyama et al. 1988). It should be pointed out that such screening effects are closely related to the "effective charge" concept for heavy ion energy loss (compare also the discussion in Sect. 2.3).

Note that, in all spectra (Fig. 3.3), there is a weak structure around $E \approx 10$ eV. A similar structure has been reported recently by Hippler (1988) and has been interpreted as "electron emission from a state of the density of final states of gold, which is at an energy level of 15.6 eV above the Fermi level".

3.2 Electron Angular Distributions: Refraction of Electrons

Figure 3.4 shows a SE angular distribution from C^+ ion impact at two velocities ($v_P = 2.30 v_B$, top, and $v_P = 2.85 v_B$, bottom). The relative intensity of SE, $N_e(\Theta)$, integrated over the SE energy interval 3 eV $\leq E_e \leq 8$ eV, is plotted here as a function of the observation angle Θ. The carbon foil targets were tilted with respect to the beam axis ($\delta = 45°$). Clearly, one can observe a peak in each distribution superimposed on the continuous "true" secondary electron background. The intensity of the low-energy "true" SE decreases strongly with increasing observation angle Θ. The peaks in the

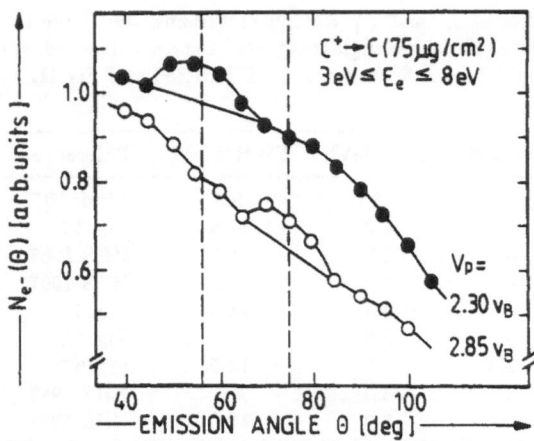

Fig. 3.4. Secondary electron angular distributions $N_e(\Theta)$ of the collision systems indicated in the figure, electron energy interval $3\,\mathrm{eV} \leq E_e \leq 8\,\mathrm{eV}$. The solid lines are drawn to guide the eye, the dashed lines indicate the mean shock electron emission angle Θ_{em}. v_B is the Bohr velocity (from Rothard et al. 1990c)

angular distributions belong to the directed emission of shock electrons perpendicular to the cone of the ion-induced wake (Echenique, Ritchie and Brandt 1979) in the electron plasma of the solid (Schäfer et al. 1978, 1980, Frischkorn et al. 1980, 1981, Brice and Sigmund 1980, Burkhard et al. 1987a, Rothard et al. 1987a, 1989a, 1990c, 1991b).

It was predicted (Schäfer et al. 1978, 1980, Brice and Sigmund 1980) that the mean emission angle of shock electrons Θ_{em}^{th} should follow the Mach-relation

$$\Theta_{em}^{th} = \arccos v_s / v_P \,. \tag{3.2}$$

Here, v_s is the shock wave group velocity, mostly depending on the plasma frequency ω_P of the solid. In Fig. 3.5, the experimentally observed mean emission angles Θ_{em} (full circles) taken from Burkhard et al. (1987a) are plotted as a function of the projectile velocity v_P. For comparison, the theoretically predicted Θ_{em}^{th} (dashed line) is included in the figure. The systematic deviation of $\Delta\Theta \approx 10°$ between the experimental data and the theoretical prediction has recently been explained in terms of low energy electron refraction at the solid surface (Rothard et al. 1989a, 1990c, 1991b).

The inset in Fig. 3.5 shows how shock electrons moving through the solid in a direction perpendicular to the wake shock front are refracted at the surface: their velocity component perpendicular to the surface is reduced corresponding to the work-function Φ. The observable mean emission angle Θ_{em}^{exp} is then given by

$$\Theta_{em}^{exp} = \delta + \arcsin[(1 + \Phi/E_s)^{1/2} \sin(\Theta_{em}^{th} - \delta)] \,. \tag{3.3}$$

The observed emission angles Θ_{em} (full circles) are well described by (3.3). Thus, shock electrons represent a unique internal source of directed low-energy electrons and can be used to study directly the refraction of low-energy electrons at solid surfaces. First studies of secondary electron an-

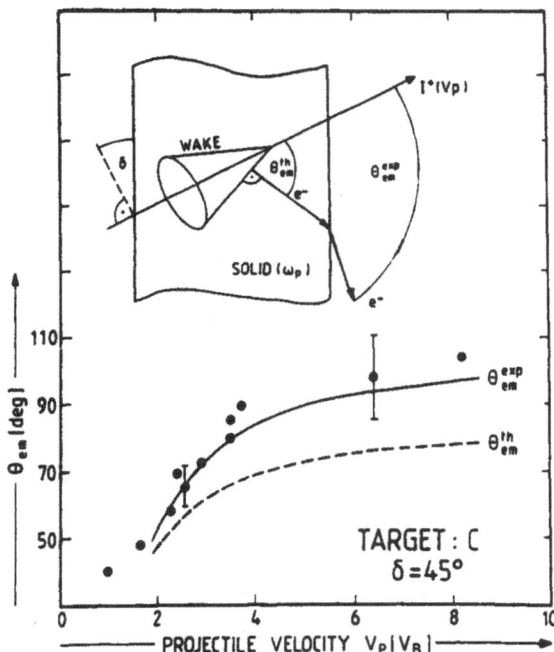

Fig. 3.5. Mean shock electron emission angle Θ_{em} as a function of the projectile velocity v_P. Full circles: experimental data, dashed line: theoretical prediction of the emission angle Θ_{em}^{th} neglecting the influence of the surface potential barrier, solid line: corrected theoretical prediction taking into account the refraction of low energy electrons at the surface for Θ_{em}^{exp}, from (3.3). The refraction phenomenon is illustrated by the inset, see text (target: carbon, tilt angle: $\delta = 45°$, mean shock electron energy $E_s \approx 5\,\text{eV}$) (according to Rothard et al. 1989a)

gular distributions from C^+-impact on single crystals of $YBa_2Cu_3O_7$ and $EuBa_2Cu_3O_7$ high T_c superconductors show evidence for shock electron emission from the excitation of the superconductors at two collective resonances $\hbar\omega_P^{FC} \approx 1.5\,\text{eV}$ and $\hbar\omega_P^{VP} \approx 25\,\text{eV}$ (Rothard et al. 1991b).

Surprisingly, the intensity $N(\Theta)$ of the low energy "true" SE background in forward direction does not strictly follow the theoretically expected (see, e. g., Sigmund and Tougaard 1981, Schou 1988) and (in backward direction) experimentally observed (Mitschler et al. 1984) cosine-dependence $N(E) \sim \cos\Theta$. This can be seen from a closer inspection of the spectra for H^+ and H_2^+ impact shown in Fig. 3.2 ($\delta = 0°$) as well as in the angular distributions from C^+ impact (Fig. 3.4, $\delta = 45°$). This may be explained by an anisotropic internal distribution of SE due to contributions from fast δ-electrons. An enhancement of Auger electron yields in forward direction from fast heavy ion impact compared to Auger yields in backward direction from thin foils has been attributed to additional ionization by fast δ-electrons (see, e. g., Schiewitz et al. 1990). The angular distribution of these Auger electrons roughly exhibits a cosine-dependence.

Furthermore, it is interesting to note that both theoretical and experimental studies (Burgdörfer, Wang, Müller 1989, Yamazaki et al. 1990) show evidence for an additional "wake"-related mechanism for electron emission: forward electron spectra ($\Theta = 0°$) induced by the penetration of anti-protons through thin carbon foils show structures which can be attributed to the emission of *wake-riding electrons* (Neelavathi, Ritchie and Brandt 1974). Such electrons originate from bound states in the wake potential of the anti-proton and can be observed as bumps at electron velocities slightly below the projectile velocity. In contrast to the case of positively charged particles, there is no convoy electron peak at $v_e \approx v_P$, and the *"anti-cusp"* caused by the repulsive interaction between the negatively charged anti-proton and the electrons is filled up by scattered electrons (Burgdörfer, Wang, Müller 1989, Yamazaki et al. 1990).

3.3 Dependence of Electron Spectra on the Surface Coverage with Adsorbates

Since the shape of SE spectra strongly depends on the contamination of the surface with adsorbates such as hydrocarbons or water, meaningful measurements must be performed in ultrahigh vacuum (UHV, $p < 10^{-7}$ Pa) with controlled surface conditions. The first evidence for a strong dependence of CE emission on surface properties was obtained from thin foils produced by standard evaporation techniques and thus contaminated with $\approx$1–2 monolayers (ML) of adsorbed hydrocarbons which subsequently were sputter-cleaned with noble gas ions (Burkhard et al. 1987b, Rothard et al. 1987b). The target surfaces were controlled simultaneously, i.e. during the ion exposure of the experiment, by a variety of independent methods such as Auger electron spectroscopy (AES), secondary electron spectroscopy (SES), Rutherford backward/forward scattering spectroscopy (RBS/RFS) and elastic recoil detection (ERD) (see, e.g., Burkhard et al. 1988a, Lorenzen et al. 1989). The residual coverage B with carbon and oxygen atoms was estimated to be lower than $B(C) < 0.2$ ML and $B(O) < 0.1$ ML (Lorenzen et al. 1989, Rothard et al. 1991a).

The strong dependence of the shape of SE spectra on the surface coverage is demonstrated in Fig. 3.6. Except for the high-energy binary encounter electrons at $E > 700$ eV, originating from close collisions between the projectile and target electrons, all other distinct structures change drastically ($> 30\%$) after sputter-cleaning of the surfaces. The (fast) binary encounter electrons mainly originate from depths of several hundred Å and thus are not affected strongly by changes of the near-surface layers.

After surface cleaning from carbon adsorbates, no carbon Auger electrons (at $E = 265$ eV) can be observed, whereas the copper MVV Auger electrons (at $E = 63$ eV) appear from the clean copper surface. The maximum residual coverage can be estimated to be $B(C) < 0.2$ ML and $B(O) < 0.1$ ML

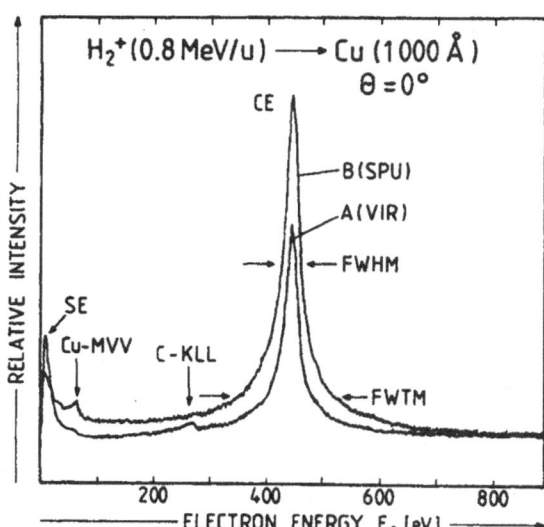

Fig. 3.6. Doubly differential SE spectra ($\Theta = 0°$, not corrected for the transmission of the spectrometer) $E\,d^2n/dE\,d\Omega$ as a function of the electron energy v_e from H_2^+ (1.6 MeV) penetration through a Cu foil (d = 1000 Å). The curve labeled A(VIR) belongs to untreated foils, and curve labeled B(SPU) to sputter-cleaned foil surfaces (from Rothard et al. 1987b)

(Rothard et al. 1991a, Lorenzen et al. 1989). The thickness of the carbon adsorbate layer without cleaning amounts to $B(C) \approx 1$ to 2 ML.

Furthermore, the height of the low-energy "true" SE distribution decreases and its FWHM increases. Generally, the energy $E_{\max}$ of the "true" SE peak maximum is shifted to higher energies: $E_{\max}$("dirty" surface) $< E_{\max}$("clean" surface). This behavior is explained by the following arguments:

1) A layer of adsorbates on a clean metal surface can lead to a reduction of the electron work-function Φ. This leads to an enhanced surface transmission probability, in particular for low-energy SE (Hölzl and Schulte 1979).

2) Layers of adsorbates or oxides lead to a larger electron escape depths λ and thus to a higher SE yield (Hasselkamp 1988, Schou 1988).

3) The change of the composition (i.e. the target material Z_T) of the near-surface layers affects the stopping power S and thus the production of SE.

4) It can be shown that a rough, uncleaned surface is associated with an enhanced electron escape probability compared to a smooth, planar surface (Borovsky, McComas and Barraclough 1988). The sputtering process applied here both cleans and smoothens the surfaces by preferential sputtering (Rothard et al. 1989a).

One of the most striking results of this type of experiment is that both the yield of CE and the shape (full width at half maximum, FWHM and full width at tenth maximum, FWTM) of the CE peak (E) are significantly enhanced for clean surfaces (Burkhard et al. 1987b, Rothard et al. 1987b,

1988a, 1991a). This shows that a large fraction of CE is formed at the last layers of the solid or even when the ion leaves the solid. The yield of CE is influenced by changes of the target material Z_T leading to changes of the electron capture and loss cross sections σ_C and σ_L as well as the CE transport lengths λ (Laubert et al. 1980, Lotz et al. 1986, Burkhard et al. 1988b, Rothard et al. 1988a).

A compilation of the results on the dependence of CE yields on the target material is shown in Fig. 3.7. In the case of light ions with $v_P \gg v_B$, the yield shows a dependence on the density of the nearly free conduction electrons (Burkhard et al. 1988b), whereas for low energy CE ($E \approx 10\,\text{eV}$) from slower heavy ions (Kr^+ at $v_P \approx v_B$), the yields possibly depend on the work-function Φ of the target (Rothard et al. 1988a).

Furthermore, the surface transmission probability related to the work-function Φ depends on the surface coverage B, in particular for low energy CE ($E < 50\,\text{eV}$) (Burkhard et al. 1987b, Sanchez, De Ferrariis and Suarez 1989, Rothard et al. 1987b, 1988a, 1991a, Sanchez 1989). The influence of adsorbates on $\Theta = 0°$ SE spectra has been studied by Sanchez, De Ferrariis and Suarez (1989) with another technique. An ion source was used to deposit sodium on the surface of an Al foil. In good agreement with the results shown in Fig. 3.6, a strong reduction of the CE yield and an increase of the "true" SE yield induced by H^+ (60 keV) was found for Na contaminated surfaces ($B(\text{Na}) \approx 0.15\,\text{ML}$). Unfortunately, the Al foils were not sputter-cleaned before the Na deposition, and no independent means of surface control (such as AES) has been applied.

A new technique (which in a sense combines the above-quoted techniques) has been introduced to study the dependence of SE emission on controlled surface contamination by condensing gases such as CO_2 and Xe on the cold surfaces of sputter-cleaned targets (Rothard et al. 1991a, Schos-

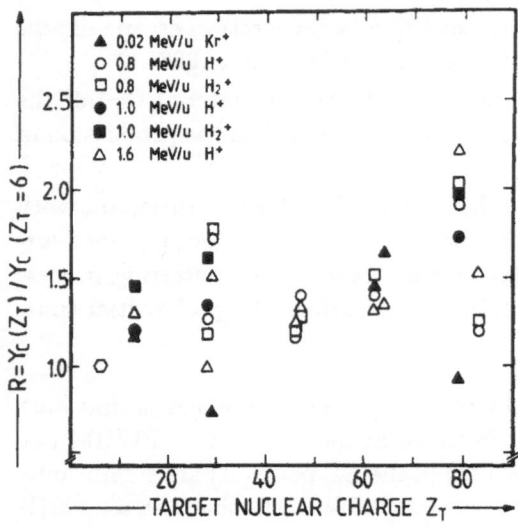

Fig. 3.7. Relative convoy electron yields $R(Z_T) = Y_C(Z_T)/Y_C(Z_T = 6)$ from sputter-cleaned foil surfaces (normalized to the yield from carbon foils) as a function of the target nuclear charge Z_T (Rothard et al. 1987b)

nig et al. 1991). The preparation of the cooled target is divided in two stages:

i) The surface is cleaned by sputtering with Ar^+ ions (0.8 MeV) using a typical ion flux of $D \sim 15\,nA\,min/mm^2$. The cleaning procedure is supervised by AES and SES (see Fig. 3.6) as well as by RBS or RFS.

ii) By condensing CO_2 on the cold, cleaned targets the surface can be contaminated in a controlled manner. Fig. 3.8 demonstrates the dependence of the SE spectra on the contamination with a frozen gas layer. The spectra from the sputter cleaned (A) and subsequently contaminated (B) copper target differ significantly:

1) The yield of convoy electrons (CE) in spectrum B is reduced as a result of CO_2 contamination, i.e. the CE yield is higher for a clean Cu surface than for a CO_2 covered surface. This result is in good agreement with the above-quoted results by Sanchez, De Ferrariis and Suarez (1989), Burkhard et al. (1987b), Rothard et al. (1987b, 1988a) and Sanchez (1989) (Fig. 3.6).

2) The Cu MVV Auger electrons of the target material at an energy of 63 eV disappeared, but a weak structure at 265 eV indicates Carbon KLL Auger electrons.

3) The low energy "true" SE are enhanced. The possible explanations (as, e.g., a reduction of the surface potential barrier Φ of the solid) have already been mentioned above in the discussion of Fig. 3.6. On the other hand, in both spectra no significant dependence of the intensity of electrons with higher energy ($E > 750\,eV$) on the surface coverage is observed, because the influence of the surface conditions on the emission of these electrons is negligible.

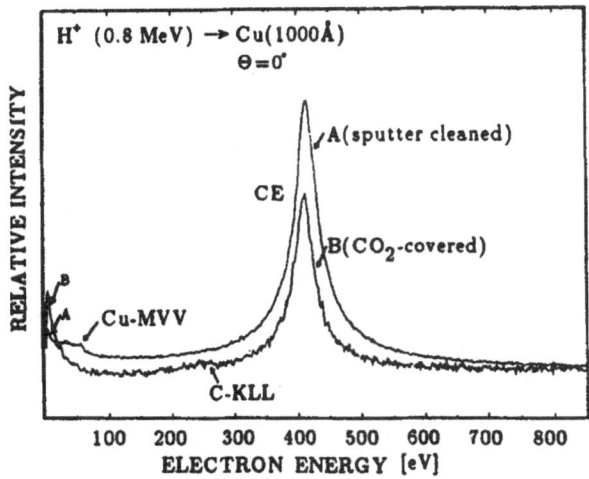

Fig. 3.8. Electron spectra ($\Theta = 0°$) from H^+ (0.8 MeV) bombardment of sputter-cleaned Cu ($d = 1000\,Å$) (A) and from Cu covered with carbon dioxide (B) (not corrected for the spectrometer transmission) (Schosnig et al. 1991, Rothard et al. 1991a)

There is experimental evidence for a target temperature dependence of the total SE yield γ_T from carbon foils Gay and Berry (1979) as well as carbon, copper and gold foils and YBaCuO high-temperature superconductors (Rothard et al. 1988b, 1990c, 1991b). A $\approx 15\%$ reduction of γ_T (temperature coefficient $\beta \approx -5^*10^{-4}\,\mathrm{K}^{-1}$), when cooling the target from $T = 300\,\mathrm{K}$ down to $T = 30\,\mathrm{K}$, was observed. First studies of $0°$ SE spectra at low temperatures (Rothard et al. 1991a) show that the low-energy part of the spectra is decreased. Furthermore, the yield of convoy electrons even at energies $E > 400\,\mathrm{eV}$ is, surprisingly, reduced.

These results may indicate that the interaction of the ions with nearly free electrons connected to collective excitations such as plasmons or the wake depends on the temperature. Also transport and surface properties have to be considered. A small (around 10%) enhancement of the binary encounter electrons around $v_e \approx 2v_P$ may be related to a small density effect. Further theoretical investigations and studies of, e. g., the T-dependence of wake-induced shock electron emission, in particular from high-temperature superconductors, may help to elucidate these findings (Griepenkerl, Müller and Greiner 1989, Rothard et al. 1991a,b).

3.4 Molecular Effects

Molecular effects in forward direction can be observed as a result of the ion charge equilibration and possibly wake effects and will be discussed in Sect. 4.2. Such effects are unlikely for thick targets because the constituents of the molecules are separated and move as independent fragments. Interestingly, in the case of convoy electrons, there is a small molecular effect: The molecular yield is increased, $R_c = Y_c(\mathrm{H}_2^+)/2Y_c(\mathrm{H}^+) \approx 1.15$ at $E_P/M_P = 1\,\mathrm{MeV/u}$ (Lotz et al. 1986). Also, the FWHM differs significantly, $\mathrm{FWHM}(\mathrm{H}_2^+) \approx 0.7\,\mathrm{FWHM}(\mathrm{H}^+)$ (Burkhard 1986). In backward direction, both the yield of low energy "true SE" and Auger electrons induced by hydrogen molecular ions at $E_P/M_P = 0.8\,\mathrm{MeV/u}$ are enhanced compared to proton impact, $R = Y(\mathrm{H}_2^+)/2Y(\mathrm{H}^+) \approx 1.2$ (Burkhard et al. 1987c, Kroneberger et al. 1988, Rothard et al. 1990c).

The binding electrons of the molecular ions can be lost near to the entrance surface due to collisions with the target atoms. These electrons, observed at $v_e \approx -v_P$ in the laboratory frame, are produced in a similar way as the binary encounter electrons from the target in forward direction observed at $v_e \approx -v_T$ in the projectile frame (with $v_T = -v_P$ being the target velocity in the projectile frame $v_P = 0$). The electron loss distribution from H_2^+ with a maximum at an electron energy E_L is deduced by subtracting the ionization electron background approximated from H^+ induced spectra (compare Fig. 3.2). Such loss electron distributions $d^2n_{\mathrm{LOSS}}/dE\,d\Omega$ are shown in Fig. 3.9 for C, Ni, Cu and Au targets. The distributions are normalized to $d^2n_{\mathrm{LOSS}}(E_L)/dE\,d\Omega = 1$.

Table 3.2. Target-material (Z_T) dependence of E_L, FWHM and I of the loss electron distributions shown in Fig. 3.9. The data[c] for the mean electron loss depths λ_L and the electron energy loss dE/dx are from Koyama et al. (1987). In the case of C, dE/dx can be estimated from Lencinas et al. 1990[b] and thus λ_L can be calculated from (3.4)[a]. No data are available for Ni

Material	Z_T	E_L [eV]	FWHM [eV]	I	λ_L [Å]	dE/dx [eV/Å*]	E_L^{calc} [eV]
C	6	300 ± 30	310 ± 25	1	45[a]	3[b]	
Ni	28	370 ± 20	245 ± 15	3.8			
Cu	29	370 ± 20	245 ± 15	4.6	10[c]	6[c]	375
Au	79	385 ± 20	255 ± 15	3.1	6.3[c]	7[c]	390

Both E_L and the FWHM of the loss electron distribution show a significant target material (Z_T) dependence. E_L is always smaller than the energy expected for a totally elastic reflection, $E_{eq} = 435\,\text{eV}$. Similar results were recently reported by Koyama et al. (1987). Table 3.2 summarizes the results for E_L, FWHM, and the relative intensity I (which may be a measure for the angular distribution). The target material dependence of E_L can be explained in terms of the energy lost by these projectile electrons (dE/dx) within a certain depth λ_L inside the solid during their transport to the surface. Within this simple model (Koyama et al. 1987) the maximum E_L^{calc} of the loss electron distribution can be calculated by

$$E_L^{calc} = E_{eq} - (\lambda_L dE/dx) . \tag{3.4}$$

A comparison of the E_L^{calc} values calculated from (3.4) to the experimental E_L data for Cu and Au shows a good agreement (Table 3.2).

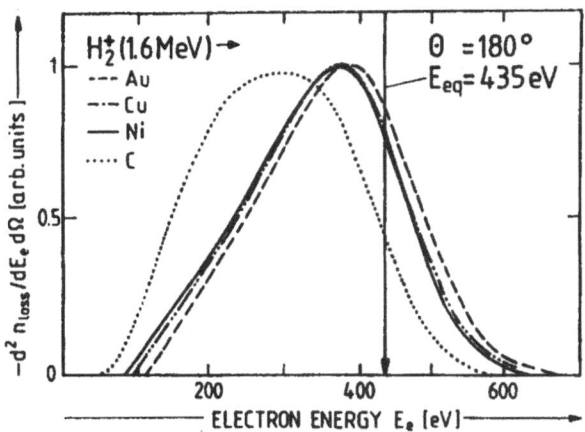

Fig. 3.9. Loss electron distributions $(\Theta = 180°)$ for several target materials (C, Ni, Cu, Au) (see text)

4. Pre-Equilibrium Electron Yields

4.1 Electron Yields as a Probe of Pre-Equilibrium Stopping Power

Most of the theoretical approaches to SE emission divide the process of SE emission into three consecutive stages (see, e. g., Sternglass 1957, Schou 1980, 1988, Sigmund and Tougaard 1981, Hasselkamp 1985, Devooght, Dubus and Dehaes 1987, Rösler and Brauer 1988, Koschar et al. 1989, Rothard et al. 1990a,b): 1) the *production* of the electron due to a single interaction at a certain point inside the solid, 2) the *transport* of the liberated electron to the surface of the solid and 3) the *transmission* through the surface potential barrier.

In the case of thin solid foil targets, and in particular if we are dealing with electron emission in the forward direction, also 4) the *preparation* of the electronic state of the projectile ion due to the dynamics of charge exchange and excitation in connection with the charge state evolution has to be considered.

The following discussion of the target-thickness (d) dependence of SE yields from thin foils within such a four-step-model is based on an extension of the semiempirical concept of Sternglass (1957) by Koschar et al. (1989). The application of a similar model to convoy electron emission will be discussed in the following chapter. It is important to note that the concept of *ion preparation* is closely connected to the concepts of *effective ion charges* and the *effective ion energy loss near the surface* (Sect. 2.3).

Following Bohr (1948), the processes leading to the ion energy loss can be considered as caused by two extremely different types of collision processes, referred to as *distant* and *close collisions*. The characteristic length that separates these two processes is the Debye screening length of the electron plasma, $\lambda_D = v_P/\omega_P$. The first process leads to direct generation of low energy SE with isotropic angular distribution around the beam axis and a mean SE energy $E_{SE} \approx 25\,\mathrm{eV}$. The total number of electrons n_{SE} per unit layer dx at a depth x (the entrance surface of the solid is at $x = 0$) is proportional to the total energy loss $dE/dx|_x$ at x divided by E_{SE}:

$$n_{SE}(x)\,dx = (1 - B)\frac{1}{E_{SE}}\frac{dE}{dx}\bigg|_x dx \ . \tag{4.1}$$

The partition factor $1 - B$ describes the contribution of distant collisions to the projectile energy loss. In the original Ansatz by Sternglass (1957) an equipartition $B = 0.5$ was chosen according to Bohr (1948).

The close collisions lead to the production of high energy δ (or "binary encounter") electrons which are emitted non-isotropically, mainly into the forward hemisphere. Their number is small ($\approx 10\,\%$) compared to the low energy SE from primary ionization, but they transport and dissipate energy on their way through the bulk and thus they initiate cascades of SE in higher order inelastic collisions far from their place of origin. The energy transport by δ-electrons can be described by an exponential diffusion function

$$f(x, \lambda_\delta) = 1 - \exp(-x/\lambda_\delta) . \tag{4.2}$$

λ_δ is a characteristic transport length comparable to the range of δ-electrons inside the solid. The resulting cascade SE are assumed to have similar angular and energy distributions as direct SE. Analogous to (4.1), their total number n_δ per dx at x is given by

$$n_\delta(x)\, dx = B \frac{1}{E_{SE}} \frac{dE}{dx}\bigg|_x f(x, \lambda_\delta)\, dx . \tag{4.3}$$

The total number of all liberated SE at the depth x inside the solid is given by the sum of (4.1) and (4.3).

The transport of SE through the bulk towards the entrance ($x = 0$) or exit surface ($x = d$) of the solid can be described by the diffusion functions

$$\begin{aligned} P_B(x) &= \tau A \exp[-x/\lambda_{SE}] \\ P_F(x) &= \tau A \exp[-(d-x)/\lambda_{SE}] . \end{aligned} \tag{4.4}$$

In a simplifying way, the coefficient A mainly contains the initial angular distribution of SE, and τ represents the surface transmission probability, both depending almost only on target properties. The diffusion length λ_{SE} corresponds approximately to the escape depth of low energy electrons ($E < E_{SE}$).

If production and diffusion of SE are considered as independent processes, the forward (γ_F) and backward (γ_B) SE yields, as function of the target thickness d, will be given by integrating the product of the production terms (4.1) and (4.3) and the corresponding diffusion probabilities (4.4):

$$\gamma_{F/B}(d) = \int_0^d [n_{SE}(x) + n_\delta(x)] P_{F/B}(x)\, dx . \tag{4.5}$$

In the high velocity limit $v_P > Z_P^{2/3} v_B$, with $\lambda_\delta \gg \lambda_{SE}$ and the ion energy loss slowly varying compared to λ_δ and λ_{SE}, the resulting electron yields are given by

$$\begin{aligned} \gamma_B(d) &= A \frac{dE}{dx}\bigg|_0 \left\{ 1 - B - (1-B) \exp\left(-\frac{d}{\lambda_{SE}}\right) \right\} \\ \gamma_F(d) &= A \frac{dE}{dx}\bigg|_d \left\{ 1 - B \exp\left(-\frac{d}{L_\delta}\right) - (1-B) \exp\left(-\frac{d}{\lambda_{SE}}\right) \right\} \end{aligned} \tag{4.6}$$

with the "material parameter"

$$A = \frac{\tau A \lambda_{SE}}{E_{SE}} . \tag{4.7}$$

123

The backward SE yield γ_B equilibrates within some SE diffusion lengths λ_{SE} to

$$\gamma_B = \Lambda(1-B)\frac{dE}{dx}\bigg|_0 \sim q_i^2 \sim S_{eB}^* \qquad (4.8)$$

and shows the proportionality of γ_B with the mean energy loss $dE/dx|_0$ at the entrance of the solid, i. e. the near-surface energy loss S_{eB}^* introduced in Sect. 2.2, and, thus, with the pre-equilibrium stopping power of the projectile ion. In contrast to γ_B, the forward SE yield γ_F mainly increases as function of λ_δ and saturates to

$$\gamma_F = \Lambda\frac{dE}{dx}\bigg|_\infty \sim q_d^2 \sim S_{eF}^* \qquad (4.9)$$

proportional to the energy loss of the projectile ion in the charge equilibrium.

Without the restriction $\lambda_\delta \gg \lambda_{SE}$, i. e. at lower projectile velocities $v_P < Z_P^{2/3} v_B$ the result is

$$\gamma_B = \Lambda^* \left(1 - \frac{B}{1+(\lambda_{SE}/\lambda_\delta)}\right)\frac{dE}{dx}\bigg|_0 . \qquad (4.10)$$

From (4.9) and (4.10), the ratio $R = \gamma_F/\gamma_B$ can be deduced (compare Sect. 2.2, (2.5)).

Experimental studies confirm the Koschar et al. (1989) model (4.5–10) concerning the velocity-dependence of the electron yields γ and of the ratio R (Chap. 2) and, in particular, concerning the target thickness ($d \cong \varrho x$) dependence of SE yields: Fig. 4.1 shows γ_F (top) and γ_B (bottom) from C^{q+} (1 MeV/u) impact on thin carbon foils. The most striking result is that γ_B is strongly related to the ion charge state $q_i = q$, whereas γ_F shows no significant dependence on q within the investigated ϱx-interval. Furthermore, γ_F increases more weakly as function of ϱx than γ_B, although γ_B does not increase as fast as expected from (4.6). A least square fit of (4.6) to the experimental data is overlayed in Fig. 4.1 for comparison.

An important result of the Koschar et al. (1989) model is the experimental value of $\lambda_\delta = 300 \pm 26$ Å for C^{q+} (1 MeV/u) and $\lambda_\delta = 1200 \pm 50$ Å for Si^{q+} (3.9 MeV/u) for the high energy δ-electrons. In contrast to λ_δ, the escape depth of low energy SE remains constant, i. e. $L_{SE} = 14 \pm 2$ Å for both projectiles within the experimental errors. This is the most evident proof for the independence between the production and the diffusion of SE inside the solid. Close to the stripping criterion $v_P = Z_P^{2/3} v_B$, which is near the stopping-power maximum, one has to expect deviations from the equipartition rule mainly due to the additional screening of tightly bound projectile electrons inside the target. This might be indicated by the slightly increased partition factor $B = 0.59 \pm 0.05$.

The target thickness dependence of γ_F and γ_B is scaled by the SE escape depth λ_{SE} and the energy transport length of δ-electrons, λ_δ. These lengths

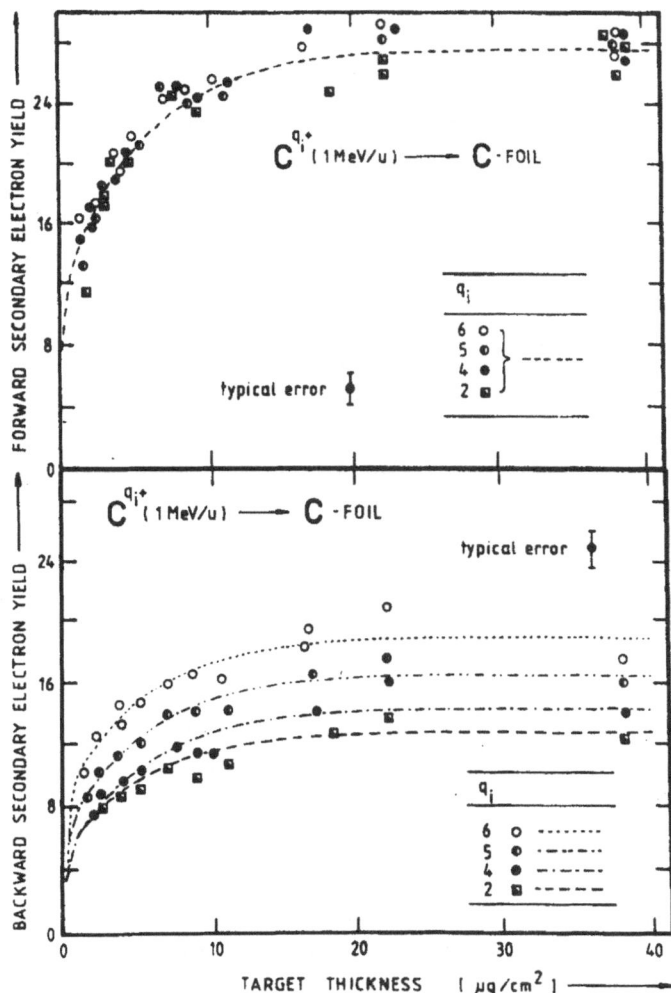

Fig. 4.1. SE yields from the forward (top) and backward surface (bottom) of carbon foils induced by C^{q+} (1 MeV/u), $q_i = 2, 4, 5, 6$ as a function of the target thickness ϱx. The lines represent least-square fits to the data with equations (4.6) (from Koschar et al. 1989)

are in good agreement with the assumptions made in the Koschar et al. (1989) model, i.e. λ_{SE} is in the order of the inelastic mean free path of "slow" (25 eV) electrons, and λ_δ is comparable to the range of δ-electrons. The results shown in Fig. 4.1 confirm the proportionality between the SE yields and the effective electronic stopping powers $S^*_{eF,B}$ at the entrance and exit both in the projectile charge equilibrium and, especially, in the projectile charge pre-equilibrium regime (compare Chap. 2). Furthermore, the measurement of γ_B as a function of the initial charge state q_i yields direct information on the pre-equilibrium effective charge of the projectiles and, thus, the shielding of bound projectile electrons at the entrance of

125

the solid. Koschar et al. (1989) reported the first evidence for shell effects concerning $S_{eB}^*(q) \sim q_i$ (Fig. 4.1). Thus, the measurement of SE yields from thin foils in forward and backward direction may offer a possibility to deduce SE-related signals, as e.g. both equilibrium electronic stopping power S_e and pre-equilibrium stopping power $S_e^*(q_i)$ near the solid surfaces (Rothard, Schou, Groeneveld 1991).

4.2 Electron Emission from Molecular Ion Impact

With molecules or clusters as projectiles, the distinction between close and distant collisions also plays an important role when studying the question, in which cases the atomic constituents of the molecule act like single, uncorrelated particles and in which cases they have to be regarded as a united charge when interacting with the target. The answer will depend strongly on the impact parameter b and the internuclear separation r_x between the atomic constituents. Molecular effects can be observed in the ratio between a physical quantity measured with molecular projectiles and the sum of the values measured with its atomic constituents, i.e. for SE yields

$$R(\gamma) = \frac{\gamma(\text{Molecular Ion})}{\Sigma\gamma(\text{Atomic Constituents})} . \qquad (4.11)$$

This ratio R should not be confused with the forward to backward electron yield ratio $R = \gamma_F/\gamma_B$. $R \neq 1$ signifies the appearance of molecular effects, which can originate from the following processes:

1) the binding electron(s) can screen the projectile charge and thus decrease the SE yield of the molecule, $(R < 1)$,

2) when the binding electron(s) are lost near the entrance surface, they can also contribute to the SE yield in addition to the yields from the atomic constituents of the molecule $(R > 1)$,

3) if $r_x < \lambda_D$, the charges of the atomic constituents act as a united charge with regard to distant collisions. Depending on the strength of the screening of the projectile charge by the electron plasma, this united charge can be larger or smaller than the sum of the charges of the atomic constituents. Furthermore,

4) the wake potential (Etchenique, Brandt and Ritchie 1979) of a leading particle may affect the energy loss of the trailing particle. Also, the superimposed wake potentials of the atomic constituents may influence the energy loss of the molecular projectile as a whole (Brandt, Ratkowski and Ritchie 1974, Brandt and Ritchie 1976, Basbas and Ritchie 1982, Kemmler et al. 1985). The internuclear separation between the atomic constituents on their way through the solid increases due to the Coulomb explosion and modified multiple scattering. Thus, the molecular effects and the contribution of the binding electron(s) will decrease in forward direction with increasing target thickness. A detailed discussion of recent results on molecular and cluster

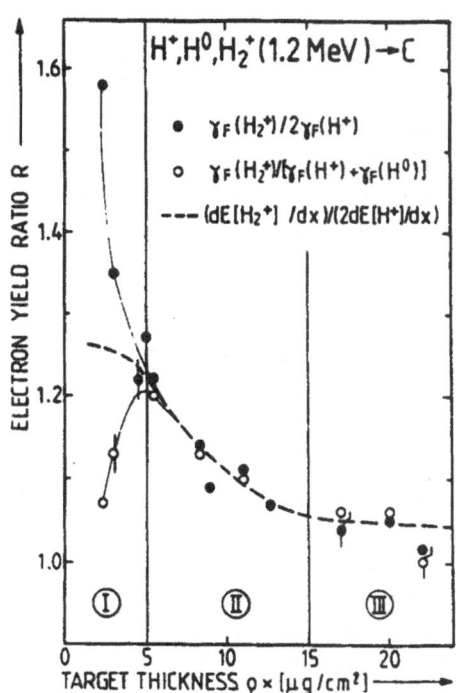

Fig. 4.2. Molecular effect ratio $R(\gamma_F)$ as indicated in the figure for H^+, H^0, and H_2^+ (1.2 MeV/u) traversing C foils (2 to 25 $\mu g/cm^2$) in comparison to the energy loss of H_2^+ (Monte-Carlo calculation) divided by twice the energy loss of a proton (dashed line, see text) (Kroneberger et al. 1988, 1989)

Within the figure:

$H^+, H^0, H_2^+ (1.2\,MeV) \rightarrow C$

- $\gamma_F(H_2^+)/2\gamma_F(H^+)$
- $\gamma_F(H_2^+)/[\gamma_F(H^+)+\gamma_F(H^0)]$
- --- $(dE[H_2^+]/dx)/(2dE[H^+]/dx)$

ELECTRON YIELD RATIO R

TARGET THICKNESS $\varrho \times [\mu g/cm^2]$

ion-induced SE emission can be found in Kroneberger et al. (1988, 1989) and Rothard et al. (1990b).

Figure 4.2 presents the "molecular effect ratio" for the forward SE yields, $R(\gamma_f)$, from H^+, H^0, H_2^+ (1.2 MeV/u) projectiles penetrating carbon foils of the specific thickness $2\mu g/cm^2 < \varrho x < 25\mu g/cm^2$ (data from Kroneberger et al. 1988). The empty circles represent the ratio taking into account the additional yield of the binding electron, i. e. $R = \gamma_B(H_2^+)/[\gamma_B(H^+) + \gamma_B(H^0)]$. The full circles represent only the sum of the yields of two independent protons. Since $R = \gamma_B(H_2^+)/[\gamma_B(H^+) + \gamma_B(H^0)]$ is close to unity, the backward molecular effect is caused mainly by the binding electron, which contributes to γ_B with about 0.6 electrons (Hölzl and Jacobi 1969). In forward direction, however, the binding electron can contribute to a molecular effect only if the target thickness is smaller than the mean range of the fast electrons, i. e. for small target thicknesses ($x < 250\,\text{Å}$, $\varrho x < 5\mu g/cm^2$, region I). Note that this thickness is in good agreement with the mean δ-electron range $\lambda_\delta = 300\,\text{Å}$ for heavy ion- induced electron yields at $E_P/M_P = 1\,MeV/u$!

In region II ($5\mu g/cm^2 < \varrho x < 15\mu g/cm^2$), the molecular effect in an order of $R(\gamma_f) \approx 1.2$ is caused by the correlated motion of the two protons through the solid. At a larger target thickness (region III, $\varrho x > 15\mu g/cm^2$), the two protons are separated and thus $R \approx 1$. A Monte-Carlo calculation of the energy loss of the proton di-cluster in the last 50 Å of the foil according to Kemmler et al. (1985) is also shown in Fig. 4.2. The ratio between this calculated energy loss and the energy loss of two independent protons is in good agreement with the measured SE yield ratio in region II and III. For a very small target thickness of $\varrho x < 2\mu g/cm^2$, it reaches a constant value of $R(dE/dx) \approx 1.26$. Energy loss calculations by Brandt and Ritchie (1976) yield the same energy loss ratio, $R(dE/dx) = 1.28$, for proton di-clusters which are not yet dissociated due to Coulomb explosion or multiscattering. These results support the assumption of a proportionality between energy loss and SE yields not only for monoatomic, but also for molecular projectiles, as well as the validity of the vicinage energy loss model for clusters (Brandt, Ratkowski and Ritchie 1974, Brandt and Ritchie 1976, Basbas and Ritchie 1982, Kemmler et al. 1985).

In Fig. 4.3 the ratios $R(\gamma_F)$ (left) and $R(\gamma_B)$ (right) from H_2^+ impact (top) on "thick" C, Al, Ti, Ni and Cu targets ($d > 200$ Å) and CO^+ impact (C and Al targets, bottom) are plotted versus the specific projectile energy (from Rothard et al. 1990b). The dashed lines indicate $R = 1$, the "no molecular effect" value. For H_2^+, the forward ratio is slightly below unity, i.e. $R = 0.85 \pm 0.15$, and shows no significant dependence on the projectile energy within the experimental errors. Since no molecular effect is expected for γ_F, this result is compatible with a value of $R(\gamma_F) = 1$. In backward direction, however, a slight increase of $R(\gamma_F)$ with E_P can be seen, starting with $R \approx 0.75 \pm 0.15$ (< 1) at 0.3 MeV/u and ending with $R \approx 1.2 \pm 0.15$ (> 1) at 1.2 MeV/u for most target materials. A reduction of the SE yield per proton for hydrogen clusters H_n^+ ($n = 2, 3$) has been observed at lower velocities $v_P < 5v_0$ (Rothard et al. 1990b).

The backward SE yields produced by CO^+ in the velocity region $v_0 < v_P < 2v_B$ with CO^+ projectiles (Fig. 4.3, bottom) are close to unity, i.e. $R(\gamma_B) \approx 0.9$, at $v_P > 1.3v_B$ for C foils, and at $v_P > 1.1v_B$ for Al foils. These velocities reflect the strength of the screening depending on the plasma

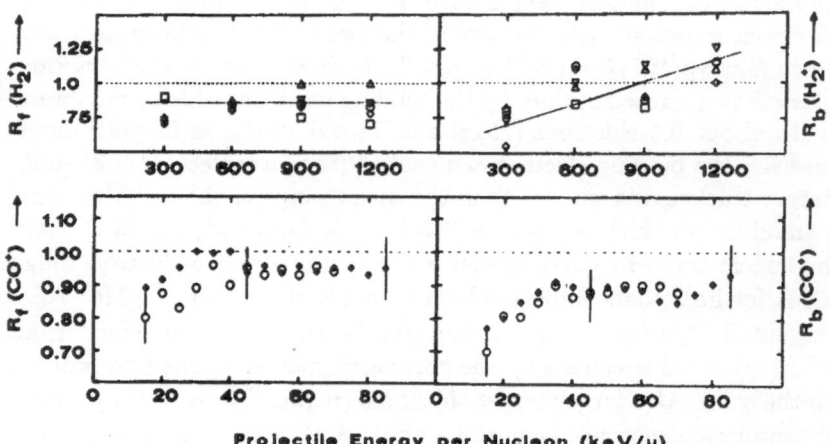

Fig. 4.3. Molecular effect ratio R for γ_F (*left*) and γ_B (*right*) from C and Al foils (open circles, C, full circles, Al) induced by CO^+ (*bottom*), and from C, Al, Ti, Ni and Cu foils induced by H_2^+ (*top*) (Kroneberger et al. 1989, Rothard et al. 1990b)

frequency in each target material being higher in C foils ($\hbar\omega_P \approx 21\,\text{eV}$) than in Al foils ($\hbar\omega_P \approx 15\,\text{eV}$). At $v_P < v_B$, the backward SE yield ratio is in the order of $R(\gamma_B) \approx 0.7$ for targets of different thickness $d > 180\,\text{Å}$. The forward yield ratios are close to unity at $E_P/M_P > 20\,\text{keV/u}$. There may be a molecular ion yield reduction $R(\gamma_F) < 1$ at lower velocities. During the passage through the foil, a (small) fraction of the molecular ions may align their axis along the beam axis due to the wake effect and suffer smaller energy losses and thus stay together for rather long distances (some $10\,\text{nm}$), in contrast to unaligned clusters (Steuer et al. 1983). The appearance of molecular effects for γ_f also for thick foils indicates that some not yet fully understood processes may play a role for SE emission.

The velocity dependence of the molecular to atomic SE yield ratios R is similar to the cluster energy loss ratio calculations by Brandt and Ritchie (1976). A slight increase of R with v_P is expected from the theory, and is confirmed by the experimental results. At lower projectile velocities, a yield reduction $R < 1$ which is not included in present theories is observed. The similar velocity dependence of the measured molecular-to-atomic SE yield ratios and the energy loss ratio calculations indicate an approximate proportionality between γ and energy loss, also for molecular projectiles at velocities around and above the Bohr velocity.

In previous experiments, where only the total SE yield γ_T from carbon foils was measured (Frischkorn et al. 1982, Frischkorn and Groeneveld 1983) with CO^+ projectiles, an oscillatory behavior for $R(\gamma_T)$ as a function of the ratio r_x/λ_w was found (r_x is the internuclear distance and λ_w is the wake wavelength). The oscillations were explained as resulting from the superposition of the wake potentials of the two atomic constituents when leaving the foil surface, i.e. the oscillations should be more pronounced in $R(\gamma_F)$. Recently, it has been mentioned by Arista et al. (1986) that the R_T results can possibly be caused by cluster vicinage effects. Within the uncertainty limits, no oscillatory behavior of R_F for carbon foils has been observed (Kroneberger et al. 1989, Rothard et al. 1990b). Data obtained with aluminum indicate that an oscillation may be present for very thin targets ($d < 200\,\text{Å}$) (Rothard et al. 1990b).

5. Pre-Equilibrium Electron Spectra: Convoy Electrons

Until today, low energy SE spectra ($E < 50\,\text{eV}$) have not been studied as a function of the target thickness because of the experimental difficulties connected with the preparation of clean surfaces. However, doubly differential electron velocity distributions at higher electron energies have been investigated in particular concerning *convoy electron* emission and the correlation to the charge state evolution of the ions. The results can be described within an unified four-step-model both for light ions and fast heavy ions (Koschar et al. 1987a,b, Kemmler et al. 1987, 1988a, Kemmler 1988, 1990).

Convoy electrons are an almost monoenergetic internal source of electrons. They serve as a valuable tool to study in detail the steps which lead also to SE emission. In the future, it may become possible to apply similar four-step-models to the case of the emission of low energy "true" SE. A general description of secondary electron spectra and yields based on the above mentioned multi-step concept is complicated, because the measured spectra in energy and angle are always composed of contributions from primary, i. e. direct ionization electrons carrying information about the preparation and direct production phase, and also from secondary or higher order electrons reflecting the transport, characterized by slowing down and cascade processes. This chapter will show on the example of a specific part of the total secondary electron spectrum, the "convoy electrons" (Breinig et al. 1982), how it is possible to learn about the interaction of the 4 consecutive stages in the secondary electron emission process.

5.1 The Four-Step-Model for Electron Emission: Preparation-Production-Transport-Transmission

The name "convoy electrons" marks a particular final state in the low energy projectile continuum: during the penetration of the ion through the solid charge exchange and excitation processes form an atomic charge cloud around the projectile. Some of these ionized electrons are not lost into completely free states and some are not captured into bound states but are transferred into high Rydberg or projectile centered continuum states where they accompany the projectile ion with the same speed and direction (Betz, Röschenthaler, Rothermel 1983).

In the electron spectrum these "convoy" electrons appear under an observation angle of $0°$ as a cusp-shaped peak with a mean electron velocity of $v_e = v_P$. One major difference in comparison to other sources for direct ionization electrons comes from the sharp localization of convoy electrons, which can be found in a small region in laboratory angle and velocity space. In this way convoy electrons can be used as a model case in studying the different aspects of secondary electron emission in solids not bringing on all the complications that arise when treating the complete spectrum.

In general the production of convoy electrons, which can take place at every place inside the solid, is described with the same concepts used in ion-atom interactions under single collision conditions. Naturally, some differences arise from the high collision rate encountered by a projectile ion penetrating a solid, which leads, e. g., to an abundant population of high l Rydberg states (Betz, Röschenthaler, Rothermel 1983, Yamazaki 1988b). In analogy to ion-atom collisions, convoy electrons can arise from two different origins: the *capture of a target electron into the continuum state of the projectile* ion (ECC) and the *electron loss of a projectile electron into the continuum* (ELC) (Breinig et al. 1982). The relative importance of these two processes depends on the evolution of the electronic configuration of the projectile ion on its way through the solid (*preparation*). A characteristic value is the mean free path for charge exchange λ_{cc} which gives a rough estimate of the target thickness x where the ion has reached its dynamical charge equilibrium.

Once the electron is transferred into the continuum state it has to reach the surface of the solid foil in order to be detected. If the presence of the isotachic projectile ion is neglected, the *transport* of the electron can be treated in the same way as for a free electron, which means that it suffers energy loss and angular deflection. The depth from where convoy electrons can still escape is roughly determined by the mean transport length λ_c. For low energy electrons, the *transmission* through the surface potential barrier has to be considered which causes additional disturbance in the flight direction and speed (compare Chap. 3).

Generally, it is not possible to study separately the influence of all these processes on convoy electron emission from solids. However, one can find some suitable cases where an aspect predominates or another can be neglected. For high electron energies $E > 100\,\mathrm{eV}$, the influence of the potential barrier at the surface of the solid can be excluded from the considerations. To minimize the contribution from different outgoing projectile charge states, projectile ions with a small number of electrons and $\lambda_c \ll \lambda_{cc}$ should be used. Additionally, it is necessary to connect the measured convoy electron yields and distributions to the corresponding outgoing projectile ion charge states by coincidence measurements (Kemmler et al. 1988a).

To describe the charge state distribution $F(q, x)$ in a gas target as a function of charge q and target thickness x, the concept of cross section $\sigma_{j,k}$ describing the charge exchange between the ground states of charge state j and k is well established (Allison 1958). The high collision frequency in a solid target, which does not permit de-excitation between (successive) single collisions, makes a simple transfer of these concepts impossible. Differences between the solid and the gas case arise also from the screening effect of the target electrons, changing the bound state wave-functions. Here it is crucial to know if higher bound and excited states of a fast projectile ion can exist in the solid. Recent experiments have shown that the $2p$ state of fast He^+ ions penetrating thin carbon foils can exist for certain projectile velocities, whereas the $4p$ state cannot (Mazuy et al. 1988).

The intensive study of charge exchange and photon emission from light ions (Clouvas et al. 1984) and heavy ions in solids (Woods et al. 1985) showed that explicitly the contributions from all possible excited states can also be applied to ion-solid collisions to describe the charge exchange between distinct states. Woods et al. (1985) attributed the enhancement of loss cross sections for C^{5+} ions in ion-solid collisions (compared to ion-atom collisions) to the enhanced population of excited states of the projectile ion in the solid. Furthermore, it was shown that for a projectile ion with only one bound electron (like C^{5+}) the excited state population should be proportional to the number of C^{5+} ions in the ground state and to the fraction of ions in the C^{6+} charge state (Cowern et al. 1983).

Another difference between ion-atom and ion-solid collisions for the charge exchange process in heavy ions is the possible post-foil autoionization of highly excited states, leading to an enhanced outgoing charge state of the projectile ion (Betz 1972). The above-mentioned studies of the charge exchange and excitation from fast projectile ions in solids support the application of the concept of Allison (1958) which describes charge exchange with a system of differential equations:

$$\frac{dF(q,x)}{dx} = \sum_{j \neq 0} [\sigma_{q+j,q}F(q+j,x) - \sigma_{q,q+j}F(q,x)] . \tag{5.1}$$

In a simple straightforward approach, the yield of convoy electrons in coincidence with a specific charge state q arising from ELC and ECC processes can be assumed to be proportional to the according charge fractions:

$$dN_{\mathrm{ECC}}(x,q) = \alpha_c \sigma_{q,q-1}F(q)n\,dx \tag{5.2}$$

$$dN_{\mathrm{ELC}}(x,q) = \alpha_L \sigma_{q-1,q}F(q-1)n\,dx . \tag{5.3}$$

Here, $dN_{\mathrm{ECC}}(x,q)/dx$ and $dN_{\mathrm{ELC}}(x,q)/dx$ are the number of convoy electrons per thickness interval dx at the position x inside the target arising from ECC and ELC. α_L and α_c are dimensionless factors which describe the fraction of electrons reaching the projectile-centered continuum state. $\sigma_{q,q-1}$ and $\sigma_{q-1,q}$ denote the capture and loss cross sections, $F(q)$ and $F(q-1)$ the charge fractions, and n the target density. If the major part of the convoy electrons can arise from highly excited states in the case of ELC in ion-solid collisions, the factor α_L can then be replaced by the expression

$$\alpha_L \rightarrow \alpha_L[1 - \exp(-x/\lambda_{ce})] . \tag{5.4}$$

In this case, the ions are in the ground state when they enter into the target. Within the mean free path λ_{ce}, the dynamical equilibrium of excited state population is reached (Hülskötter, Burgdörfer, Sellin 1987, Kemmler et al. 1988b).

A comparison of the target-thickness dependent convoy electron yield Y_e in coincidence with a specific charge state q and the charge state distribution $F(q,x)$ itself can yield important information about the possible contributions of electron loss and/or capture (Kemmler et al. 1988b): in the case of Ni^{28+} ions emerging from carbon foils at $15.6\,\mathrm{MeV/u}$, the coincident convoy electrons originate both from ECC and ELC (from Ni^{27+} states formed in the solid by capture of target electrons) with nearly the same probability (Kemmler 1990). The contribution of both processes strongly depends on the target thickness and the projectile velocity. However, an argument against the possibility of an ECC process in solids comes from the symmetric form of convoy electron distributions measured in ion-solid collisions.

Attempts had been made to extract details of the production process by fitting the convoy electron distribution in terms of a Legendre polynomial expansion (Berry et al. 1986). Several characteristics extracted from analyzed spectra originating from ion-solid collisions resemble ELC distributions measured under single collision conditions, but other details of the structure do not resemble ELC nor ECC (Focke et al. 1986). Koschar et al. (1987a) made a difference between *direct* and *indirect ELC* (DELC, IELC) and *direct* and *indirect ECC* (DECC, IECC) if these processes take place with the projectile ion in its incident charge state ("direct") or if a charge exchange event has occurred before ("indirect").

When the convoy electron is "born" inside the solid it can be treated, in a first-order approximation, as a free electron travelling with speed v_P in a direction parallel to the projectile ion. Therefore, it can suffer energy loss in inelastic collisions with electrons inside the solid, and it will be deflected in elastic collisions with the core atoms. A more sophisticated treatment of the problem has to take into account the influence of the accompanying projectile ion charge on the transport behavior. This has been done in a recent publication by Burgdörfer and Gibbons (1990), who showed that in the case of heavy ions the isotachic projectile charge strongly influences the electron transport. Because of the inherent complexity of the problem they used a Monte-Carlo code to describe the electron transport phenomena. In this article, an analytic description of the problem will be used. Although it cannot account for the complex interaction mechanism of an electron and an isotachic projectile ion travelling through a solid, it will give a more lucid view into the underlying basic mechanism, which should also be valid for light ions.

Important parameters in describing electron transport are the mean free paths λ, defined as the inverse of the integrated corresponding interaction cross section. For electron energies from 100 to 10.000 eV, the mean free paths for inelastic and elastic scattering (λ_i and λ_s) are in the order of 10 to 100 Å. Thus, if the source of electrons is located in a depth equal or larger than this value, transport effects must be taken into account. In the case of Auger transitions, for example, the signature of electron transport is a broad low energy shoulder in the Auger electron distribution (compare Chap. 3). In the case of convoy electron distributions, several aspects of the transport phenomenon, as, e.g., i) the emittance of the source, ii) the solid angle and energy resolution of the electron spectrometer, have to be considered. Also, iii) the effect of the scattering process differs between the near peak regime, when the electron suffers only one or two collisions and the off peak regime when multiple collisions have to be taken into account.

To provide a test for the theoretical tools used in the description of electron transport in the region of several 100 eV to several 1000 eV, experiments with monoenergetic electrons have been performed by Lencinas et al. (1990). Since it is impossible to achieve single collision conditions for electron scattering at these energies, a transport calculation according to Ashley et al. (1979) could be introduced for the description of the inelastic scattering. It was shown that the inelastic and elastic scatterings were separable also in this energy region, where the mean energy loss of the electrons reaches more than 10 % of the incident energy. Thus, the multiple elastic scattering could be successfully described with a slightly modified Goudsmit and Saunderson (1940) approach. A striking result of these studies is that the delta-term describing the unscattered portion of electrons is still present up to target thickness x reaching 4 times the mean elastic free path δ_s. This effect is produced by the small solid angle of the electron spectrometer ($\Delta\Omega = 10^{-5}$ st) which detects nearly the total amount of unscattered electrons at an observation angle of $\Theta = 0°$, whereas for $\Theta \neq 0°$ only a small fraction of the electron distribution (which shows only a smooth angular dependence if $\Theta \neq 0°$) can be detected.

5.2 Application to Convoy Electron Emission

In the laboratory frame, the emission of convoy electrons is strongly forward peaked, also the width of the velocity distribution is relatively small. Considering these attributes, convoy electrons resemble free monoenergetic electrons. One important difference is that the convoy electrons are produced continuously inside the solid with a thickness-dependent intensity according to the loss and capture processes discussed above.

A very interesting case where the transition from a quasi delta-source at the entrance of the foil to a monotonous source distribution inside the solid can be studied is the collision system H^0 (3 MeV/u) $\rightarrow$ C foil (Koschar et al. 1987b, Fig. 5.1). Because of the large electron loss cross section, the major part of the bound electrons (a certain fraction of which also supplies the ELC mechanism) are lost in the first few Å of the solid. At a certain target thickness, these former convoy electrons have suffered so much energy loss that they form a well separated hump behind the sharp cusp shaped convoy electron peak. This hump resembles the Landau distribution of monoenergetic electrons suffering multiple collisions and in fact one can describe the transport of these electrons with the same methods as for monoenergetic electrons (Koschar et al. 1987b). Since the convoy electrons are consecutively produced in the bulk of the solid, the situation is different near the convoy electron peak. Here, it is necessary to apply the picture of few collisions.

A relation between the differential inelastic and elastic scattering cross sections and mean free paths λ_i and λ_s and the characteristics of the transport for convoy electrons can only be found in applying a model covering all the aspects discussed above. Two different types of information are accessible investigating convoy electron production in solids. These are the shape $J(q_i, q, v, \Theta, x)$ of the distribution and the yield $Y_e(q_i, q, x)$ i.e. the total number of convoy electrons per projectile ion. $J(q_i, q, v, \Theta, x)$ is the number

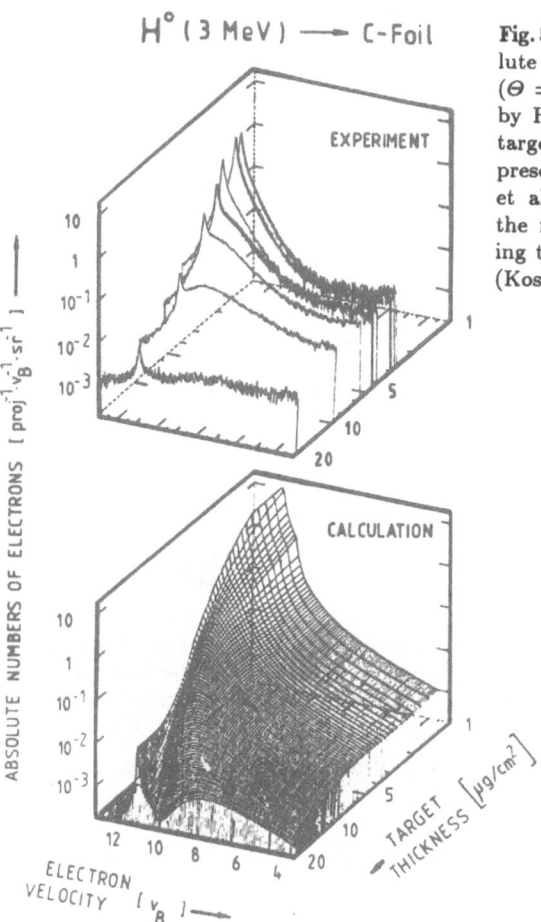

H^0 (3 MeV) ——→ C-Foil

EXPERIMENT

CALCULATION

ABSOLUTE NUMBERS OF ELECTRONS [proj^{-1}·v$_B^{-1}$·sr^{-1}]

ELECTRON VELOCITY [v$_B$]——→

TARGET THICKNESS [µg/cm²]

Fig. 5.1. Three-dimensional plot of absolute doubly differential electron spectra ($\Theta = 0°$) from carbon foils penetrated by H^0 (3 MeV/u) as a function of the target thickness ϱx. The top of the figure presents experimental results (Koschar et al. 1987b), whereas the bottom of the figure shows a calculation according to the model presented in the text (Koschar et al. 1987a,b, Kemmler 1988)

of electrons emitted after a target thickness x for an incident charge q_i and an outgoing charge q which travel with the velocity v into the direction Θ.

Both quantities are linked by

$$Y_e(q_i, q, x) = \int_{\Delta\Omega} d\Omega \cdot \int_{v_P \pm v_i} dv\, J(q_i, q, v, \Theta, x) \tag{5.5}$$

where $2v_i$ is the velocity interval and $\Delta\Omega$ is the solid angle of integration. For the calculation of $J(q_i, q, v, \Theta, x)$ the Ansatz from Tougaard and Sigmund (1982) is used. The situation is sketched in Fig. 5.2. The convoy electron source $Q(q_i, q, v', \Theta', x')$ is located in the depth x' inside the solid and emits electrons as a function of velocity v' and direction Θ'. A Legendre polynomial expansion is used to describe the doubly differential distribution (Meckbach, Nemirovsky and Garibotti 1981). The total intensity follows the charge state fraction as discussed above (5.4).

SINGLE COLLISION CONDITIONS

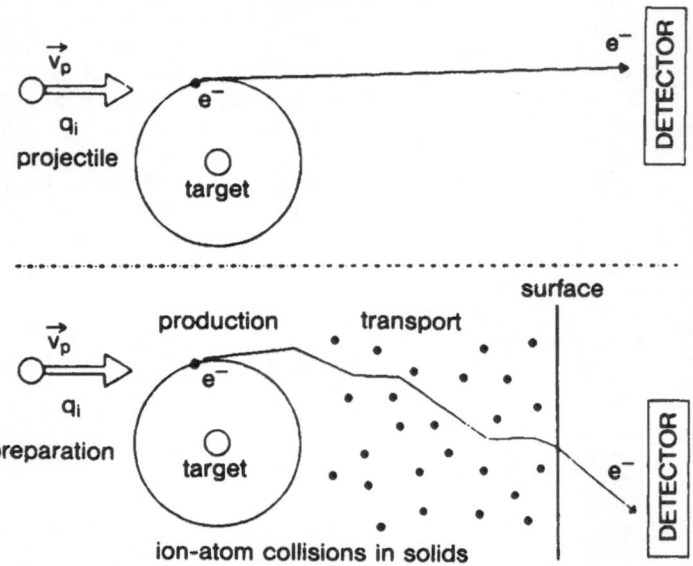

Fig. 5.2. Preparation, production, transport and transmission of convoy or secondary electrons in solids (ion-induced interaction) in comparison to the case of electron production (primary ionization) under single collision conditions (ion-atom interaction)

The distribution outside the solid is given by the expression

$$J(q_i, q, v, \Theta, x) = \int_0^x dx' \int_0^{4\pi} d\Theta' \int_0^\infty dv' Q(q_i, q, v', \Theta, x') P(v', \Theta', x'; v, \Theta) \,. \tag{5.6}$$

Here $P(v', \Theta', x'; v, \Theta)$ describes the transport of the electrons with velocity v' and direction Θ' at depth x to the surface where they escape with v and Θ. As the elastic and the inelastic scattering are separable, one can write

$$P(v', \Theta', x'; v, \Theta) = F(x', |x - x'|) f(x', |\Theta - \Theta'|) \tag{5.7}$$

with $F(x', |x - x'|)$ describing the inelastic straggling and $f(x', |\Theta - \Theta'|)$ the elastic multiple scattering. The analytical expressions for both functions are given by Lencinas et al. (1990).

To extract an expression for the convoy electron yield Y_e from (5.6), and also to obtain a relation between the mean free paths λ_i and λ_s and the transport length for convoy electrons λ_c, some assumptions have to be made. As a simple approach, the source distribution for convoy electrons can be described as a delta-function at the projectile velocity v_P

$$Q(q_i, q, v', \Theta, x') = A\delta(v' - v_P) \tag{5.8}$$

where A denotes the intensity. By assuming that only electrons which suf-

136

fered no energy loss can be detected, and neglecting angular scattering, from (5.6–8) follows

$$J(q_i, q, v, \Theta, x) = A \int_0^x dx' \int_0^{4\pi} d\Theta' \int_0^\infty dv' \delta(v' - v_P) \exp(-x'/\lambda_s) \exp(-x'/\lambda_i) \,.$$

$$(5.9)$$

The mean transport length for convoy electrons λ_c is obtained from the relation

$$1/\lambda_c = 1/\lambda_\Sigma = 1/\lambda_i + 1/\lambda_s \,, \tag{5.10}$$

where $\lambda_\Sigma = \lambda_c$ can be interpreted as the extinction length for a point source (Koschar et al. 1987a).

For the forward peaked convoy electron distribution, the yield Y_e (in units of convoy electrons per incident ion) can only be obtained by integrating the double differential distribution over a certain velocity and angular interval. The application of such a procedure does not allow to neglect electrons which are scattered into a certain velocity and angular interval being still within the integration limit. When substituting (5.10) by a more realistic expression, the inelastic scattering of the electrons has to be taken into account. Because of the small solid angle, the elastic contribution can be understood as a yes/no process (Kemmler et al. 1988c, Kemmler 1988). The inelastic scattering can be treated by an expansion into an increasing number of scattering events. With the differential inelastic cross section $d\sigma_i/dT$ and a maximum energy loss ε_i of the electron not exceeding the related velocity integration interval v_i, one can define a mean free path λ_x for those loss events as

$$\frac{1}{\lambda_x} = n \int_0^{\varepsilon_i} d\sigma_i \tag{5.11}$$

which expresses the mean depth of origin of an electron which can still be detected in the velocity interval v_i. With (5.11), (5.9) transforms to

$$J(q_i, q, v, \Theta, x) = A \int_0^x dx' \int_0^{4\pi} d\Theta' \int_0^\infty dv' Q(q_i, q, v', \Theta, x')$$

$$\times \exp[-x'/(\lambda_s + \lambda_i)]\{1 + x'/\lambda_x\} \,. \tag{5.12}$$

The transport properties from (5.12) can be written as a single exponential of the form $\exp(-x'/\lambda_s)$, i. e.

$$\frac{1}{\lambda_c} = \frac{1}{\lambda_i} + \frac{1}{\lambda_s} - \frac{1}{x' \ln\left(1 + \frac{x'}{\lambda_x}\right)} \tag{5.13}$$

where the relation between the transport lengths λ_c, λ_i and λ_s is substituted by an expression which depends on the depth of origin x' of the emitted electron.

Since recent experimental results suggested strongly enhanced convoy electron transport lengths λ_c compared to the expected values for monoenergetic isotachic free electrons (Sellin et al. 1984, Schramm et al. 1985, Kemmler et al. 1988b), it is important to discuss the model for production and transport of convoy electrons in detail. In order to exclude any artificial (effect which can produce an apparently higher transport length λ_c than expected from λ_i and λ_s), the model has to describe the different aspects of the process as good as possible.

Equation (5.13) expresses the fact that λ_c extracted from a measured distribution is in general larger than it could be expected from the inelastic and elastic mean free paths. In contrast to these findings made with heavy ions, in experiments with light ions no difference between convoy and free electron transport lengths has been found (Koschar et al. 1987a,b). However, the model can also be applied to the case of light ion impact. This is demonstrated in Fig. 5.1, where a three dimensional plot of experimental (top) and calculated (bottom) $\Theta = 0°$ electron velocity spectra from H^0 penetration of C foils of different thickness ϱx are shown (Koschar et al. 1987b).

The coincident convoy electron yield $Y_e(q_i = 28, q = 28)$ and $Y_e(q_i = 28, q = 27)$ for Ni ions (15.6 MeV/u) is plotted in Fig. 5.3 as a function

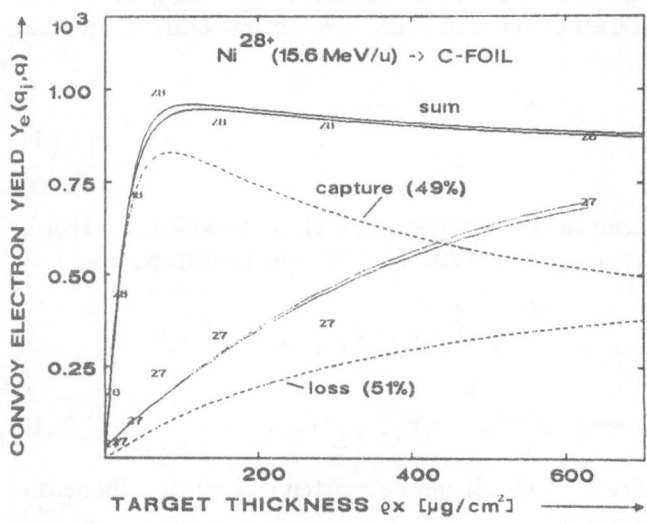

Fig. 5.3. Coincident convoy electron yield $Y_e(q_i = 28, q_f = 28)$ and $Y_e(q_i = 28, q_f = 27)$ for Ni ions (15.6 MeV/u) as a function of target thickness. The solid lines are fits according to the model explained in the text, the dashed lines demonstrate the fractions of ECC (capture) and ELC (loss) contributing to the total convoy electron yield (Kemmler et al. 1988b)

of the target thickness ϱx (Kemmler et al. 1988b). The dashed lines show the relative importance of the ECC (capture) and ELC (loss) process. A comparison with the charge state distribution (see Kemmler et al. 1988b) demonstrates that for $Y_e(q_i = 28, q = 28)$ the equilibrium value for $\varrho x > 70\,\mu g/cm^2$ can only be explained by assuming nearly equal contributions for both processes. If convoy electrons are only produced by ELC, the yield should increase further. In the case of $Y_e(q_i = 28, q = 27)$ the situation is not so clear, because the charge state fractions responsible for ECC ($q = 27$) and for ELC ($q = 26$) both increase. Consequently, the convoy electron yield follows strictly the general increasing trend of the charge fractions which still are not in equilibrium at a target thickness of $\varrho x = 650\,\mu g/cm^2$ (just like the convoy electron yield). In this case, it is also more complicated to determine the relative contribution of ECC and ELC.

Furthermore, the effect of electron transport, which manifests itself in an exponential decrease of the yield, cannot be distinguished from the increase produced by the charge exchange processes. The change in yield caused by the transport length for convoy electrons is only visible when the source term for convoy electrons starts from a high level at thin target thicknesses. This is obviously the case for the yield dependence of $Y_e(q_i = 28, q = 28)$ caused by the large amount of convoy electrons produced by the ECC mechanism. Earlier results lead to a transport length $\lambda_c = 1200$ Å in this case if a fitting procedure neglecting the contribution from inelastic scattered electrons is used. The fit using the Ansatz from (5.12) is nearly indistinguishable from this result, but the value for the transport length is different. Under the assumption that the mean free path for resting in the covered velocity interval, λ_x, has the same length as the extinction length λ_Σ, a value of $\lambda_x = 800$ Å and $\lambda_\Sigma = 800$ Å can be obtained. The mean free path for inelastic and elastic scattering for a screened Rutherford potential is $\lambda_s = 193$ Å and within the model by Ashley et al. (1979) $\lambda_i = 151$ Å is found which leads to λ_Σ (theor.) $= 85$ Å. Within the present model for the convoy electron yield dependence on the target thickness which considers every known step of convoy electron emission (preparation of the charge state, the *production* mechanisms and the *transport* of the electrons), only an enhanced transport length for convoy electrons can explain the measured yield dependence.

An additional hint to an enhanced transport length comes from the absolute convoy electron yield $Y_f(q) = Y_e(q)/F(q)$ coincident with the number of outgoing projectile ions with charge q. From the Ni^{28+} (15.6 MeV/u) data (Fig. 5.3) a value of $Y_f = 1.6 \cdot 10^{-3}$ convoy electrons per projectile ion is obtained. The error is estimated to a factor of two because the absolute normalization is based on measurements from Koschar et al. (1887a) made with different projectile ions in a different projectile velocity region. Using the model for the equilibrium convoy electron yield from Kemmler et al. (1989a), which is slightly changed under consideration of (5.12) to

$$Y_f(q, x = \infty) = n(\lambda_\Sigma + \lambda_\Sigma^2/\lambda_x)\sigma_{q,q-1}\{\alpha_L + \alpha_c\} \qquad (5.14)$$

the relative contributions of ECC and ELC to the charge exchange process, expressed by the term $\{\alpha_L + \alpha_c\}$, can be extracted.

With the results of Kemmler et al. (1988b) and with the above-given results for the enhanced transport length we can estimate that 3.2 % of all lost and captured electrons are convoy electrons. If the mean free path would be comparable to values from isotachic free electrons with $(\lambda_\Sigma + \lambda_\Sigma^2/\lambda_x) = 141\,\text{Å}$, this number would increase to 37 %! Surprisingly, convoy electrons from very deep layers inside the solid can escape. These findings support the theoretical picture for an enhanced transport path for electrons accompanying highly charged ions which was proposed by Burgdörfer (1988).

Equation (5.12) gives evidence that in the near convoy electron peak regime a certain quantity of inelastic scattered electrons are present. To study their influence on the double differential convoy electron distribution according to (5.6), which does not a priori neglect the multiple elastic scattering, this distribution has been calculated assuming a constant source term by Kemmler et al. (1888c). Since no angular spreading is visible, these results confirm the Ansatz from (5.12). Furthermore, they demonstrate that the width of the distribution parallel to the beam direction is increased as a result of inelastic scattering, but the general shape of the distribution is conserved, in accordance with (5.12). The collision system Ni (15.6 MeV/u) $\rightarrow$ C foil gives the possibility to study the shape of the double differential convoy electron distributions both for target thicknesses $x < \lambda_\Sigma$ and $x > \lambda_\Sigma$.

The measured convoy electron distributions show generally that they are smaller in the direction $v_\parallel$ than in the direction $v_\perp$ (Kemmler et al. 1989a). This is on the one hand caused by the underlying production mechanism for the convoy electrons, i. e. ECC or ELC events, but is on the other hand a consequence of the shape of the resolution volume of the electron spectrometer. For the thicker targets the tail of the distribution is more pronounced in direction $v_\perp$ than for the thinner targets. The central area is similar in both cases. These electrons, filling the tail of the distribution with increasing target thickness as a kind of "background", arise from elastic scattering events. For N^{28+} the FWHM $\Gamma_\perp$ of the distribution shows an increase from thin to thick targets.

A further interesting quantity is the absolute value and the evolution of the FWHM $\Gamma_\parallel$ in direction $v_\parallel$. Even at $\varrho x = 3\,\mu\text{g/cm}^2$, $\Gamma_\parallel$ is still higher than the width calculated from the resolution of the spectrometer. This is partly an effect of the "real" width of ELC and ECC distributions, which are in the case of ELC about 50 % and for ECC about 20 % larger than the values calculated from the pure resolution. The distinctively larger values for $\Gamma_\parallel$ can be attributed to the fraction of inelastically scattered electrons. For incident Ni^{27+}, the change of $\Gamma_\parallel$ is not so pronounced and $\Gamma_\perp$ decreases slightly with increasing target thickness.

For small target thicknesses, the fraction of ELC is enhanced, and therefore the related distribution is broader, compared to incident Ni^{28+}. With increasing thickness, the ELC contribution decreases and compensates in this way the simultaneous increase of the distribution by inelastic scattering. However, the discrepancies in FWHM between the thinnest and thickest targets are not so strongly pronounced. The calculations by Kemmler et al. (1988c) show that the differences in the tails of the distributions are larger. More insight concerning the mechanisms leading to convoy- and secondary electron emission can be obtained by combining the techniques of controlled surface contamination (Sect. 3.3) with the coincidence measurement of electrons and charge analysed ejectiles (Chaps. 1, 5) (Kroneberger et al. 1991).

6. Conclusion

In this review, the kinetic forward and backward electron emission from ion penetration through thin foils and the special relation to the pre-equilibrium of charge state distributions has been discussed. Experiments on electron emission from *thin* foils yield very detailed and basic information, because a considerable proportion of the projectile energy loss leads to the creation of high-energy electrons, which are predominantly ejected in forward direction. Furthermore, *convoy electron* emission can be observed only from thin foils in forward direction. In particular, the thickness dependence of the electron emission gives unique data on charge pre-equilibrium penetration processes.

Experiments on electron emission from thin foils allow us to study the *evolution of pre-equilibrium- to equilibrium-conditions* of such processes as electron capture, electron loss, the charge states, the excitation states and the energy loss. A main result is the proportionality between the total electron yields and the electronic energy loss both in the projectile charge equilibrium and, especially, in the projectile charge pre-equilibrium regime. This allows us to *extend the semiempirical model of Sternglass as well as Schou's transport theory*, by assuming that the *yields are proportional to the effective electronic energy loss of the ions* near the solid surfaces. Thus, ion-induced electron emission can serve as a valuable test for theories, which can also be applied to such fields as scanning electron microscopy and all kinds of surface analysis with Auger, photo or secondary electrons.

Furthermore, electron emission may become a powerful tool to measure the energy loss of charged particles and to test energy loss models, even in cases which are not easily accessible experimentally now. In particular, the energy loss of molecules or clusters or the dependence of the energy loss on the pre-equilibrium evolution of the charge state of the ions can be investigated. With sufficiently thin targets or at sufficiently high ion velocities (which are accessible with heavy ion accelerators such as GANIL or GSI–SIS) we are generally dealing with *pre-equilibrium conditions* where stopping power values can significantly differ from equilibrium data and are neither available in the literature nor easily accessible experimentally.

In general, electron emission can be described by four-step-models which divide the electron emission process into four phases: 1) Preparation of the charge and excitation state of the ions. 2) Production of electrons (in relation to the energy loss of the projectiles). 3) Transport of the electrons towards the surface and 4) transmission through the surface. This novel approach gives new insight in the fundamental processes of ion penetration through solids and gives needed heavy ion stopping power data for operation and experiment with the new generation of high energy heavy ion accelerators. Experiments on the influence of the target temperature and, in particular, *controlled surface contamination* on electron emission (which can, e. g., be studied by *freezing gases such as CO_2 or Xe on the sputter-cleaned surfaces of thin metal foils* at low temperatures) can yield important information and may be the key to use ion-induced electron emission as a valuable tool for materials and surface analysis.

Also, the fingerprints of collective excitations in electron spectra and angular distributions (*plasmon decay* and *shock electrons*) can be used to study the electronic properties (as, e. g., the conduction electron density) of solids (and even of the new high T_c superconductors). Since shock electrons are a *unique internal source of directed low energy electrons*, they can be used to study such interesting phenomena as the *refraction of low energy electrons at solid surfaces*. Furthermore, the emission of *low energy convoy*- and *"true secondary"* electrons can be used to study surface properties.

Acknowledgement. We are grateful to our collaborators C. Biedermann, A. Clouvas, N. Keller, P. Koschar, K. Kroneberger, S. Lencinas, P. Lorenzen, M. Schosnig and E. Veje. We also thank J. Burgdörfer, J. C. Dehaes, P. M. Echenique, W. Greiner, M. Inokuti, W. Meckbach, R. H. Ritchie, P. Sigmund, H. Stöcker and J. Schou for important discussions and helpful suggestions. This work has been funded by the German "Bundesminister für Forschung und Technologie" (BMFT), Bonn, Germany, under contract number 06 OF 110/II Ti 476.

References

Allison S. K.1958: Rev. Mod. Phys. **30**, 1137

Arista, N. R., Jakas, M. M., Lantschner, G. .H., Eckardt, J. C. 1986: Phys. Rev. A **34**, 5112

Arnau, A., Peñalba, M., Echenique, P. M., Flores, F., Ritchie, R. H. 1990: Phys. Rev. Lett. **65**, 1024

Ashley, J. C., Cowan, J. J., Ritchie, R. H., Anderson, V. E., Hölzl, J. 1979: Thin Solid Films **60**, 361

Basbas, G., Ritchie, R. H. 1982: Phys. Rev. A **25**, 1943

Biedermann, C., Kemmler, J., Rothard, H., Burkhard, M., Heil. O., Koschar, P., Kroneberger, K., Groeneveld, K. O. 1988: Physica Scripta **37**, 27

Berry, S. D., Elston, S. B., Sellin, I. A., Breinig, M., DeSerio, R., Gonzalez-Lepera, C. E., Liljeby, L. 1986: J. Phys. B **19**, L149

Betz, H. D. (1972): Rev. Mod. Phys. **44**, 465

Betz, H. D., Röschenthaler, D., Rothermel, J. 1983: Phys. Rev. Lett. **50**, 34

Bohr, N. 1948: Mat. Fys. Medd. Dan. Vid. Selsk. **18**, No. 8

Borovsky, J. E., McComas, D. J., Barraclough, B. L. 1988: Nucl. Instrum. Meth. B **30**, 191

Brandt, W., Ratkowski, A., Ritchie R. H. 1974: Phys. Rev. Lett. **33**, 1325

Brandt, W., Ritchie, R. H. 1976: Nucl. Instrum. Meth. **132**, 43

Breinig, M., Elston, S. B., Huldt, S., Liljeby, L., Vane, C. R., Berry, S. D., Glass, G. A., Schauer, M., Sellin, I. A., Alton, G. D., Datz, S., Overbury, S., Laugert, R., Suter, M. 1982: Phys. Rev. A **25**, 3015

Brice, D. K., Sigmund, P. 1980: Mat. Fys. Medd. Dan. Vid. Selsk. **40**, No. 8

Brusilovsky B. A. 1990: Appl. Phys. A **50**, 111

Burgdörfer, J. 1988: in *High-Energy Ion-Atom Collisions*, Berenyi, D., Hock, G. (Eds.), (Springer, Heidelberg), Lecture Notes in Physics **294**, 344

Burgdörfer, J., Gibbons, J. 1990: Phys. Rev. A **42**, 1206 .

Burgdörfer, J., Wang, J., Müller, J. 1989: Phys. Rev. Lett. **62**, 1599

Burkhard, M. 1986: Thesis, J. W. Goethe-Universität, Frankfurt am Main, Germany **IKF-D-380**

Burkhard, M., Rothard, H., Biedermann, C., Kemmler, J., Kroneberger, K., Koschar, P., Heil, O., Groeneveld, K. O. 1987a: Phys. Rev. Lett. **58**, 1773

Burkhard, M., Rothard, H., Biedermann, C., Kemmler, J., Koschar, P., Groeneveld, K. O. 1987b: Nucl. Instrum. Meth. B **24/25**, 143

Burkhard, M., Koschar, P., Heil, O., Kemmler, J., Köver, A., Szabo, Gy., Berenyi, D., Groeneveld, K. O. 1987c: Nucl. Instrum. Meth. B **24/25**, 189

Burkhard, M., Rothard, H., Kemmler, J., Kroneberger, K., Groeneveld, K. O. 1988a: J. Phys. D **21**, 472

Burkhard, M., Rothard, H., Kroneberger, K., Groeneveld, K. O. 1988b: Nucl. Instrum. Meth. B **34**, 166

Burkhard, M., Lotz, W., Groeneveld, K. O. 1988: J. Phys. E **21**, 759

Burkhard, M. F., Rothard, H., Groeneveld, K. O. 1988: Phys. Status Solidi (b) **147**, 589

Clerc, H. C., Gerhardt, H. J., Richter, L., Schmidt, K. H. 1973: Nucl. Instrum. Meth. **113**, 325

Clouvas, A., Gaillard, M. J., de Pinho, A. G., Poizat, J. C., Remilleux, J., Desesquelles, J. 1984: Nucl. Instrum. Meth. B **2**, 273

Clouvas, A., Rothard, H., Burkhard, M., Kroneberger, K., Biedermann, C., Kemmler, J., Groeneveld, K. O., Kirsch, R., Misaelidis, P., Katsanos, A. 1989: Phys. Rev. B **39**, 6316

Clouvas, A., Katsanos, A., Farizon-Mazuy, B., Farizon, M., Gaillard, M. J. 1991: Phys. Rev. B **43**, 2498

Cowern, N. E. B., Woods, C. H., Sofield, C. J. 1983: Nucl. Instrum. Meth. **216**, 287

Da Silveira, E. F., Jeronymo, J. M. F. 1987: Nucl. Instrum. Meth. B **24/25**, 534

Dednam, C. C., Froneman, S., Mingay, D. W., Van Waart, J. 1987: Nucl. Instrum. Meth. B **24/25**, 366

Dehaes, J. C., Carmeliet, J., Dubus, A. 1986: Nucl. Instrum. Meth. B **13**, 627

Devooght, J., Dehaes, J. C., Dubus, A., Cailler, M., Ganachaud, J. P. 1991: Springer Tracts in Modern Physics, preceding volume

Devooght, J., Dubus, A., Dehaes, J. C. 1987: Phys. Rev. B **36**, 5093

Echenique, P. M., Ritchie, R. H., Brandt, W. 1979: Phys. Rev. B **20**, 2567

Focke, B., Nemirovsky, I. B., Gonzales-Lepera, E., Meckbach, W., Sellin, I. A., Groeneveld, K. O. 1986: Nucl. Instrum. Meth. B **2**, 235

Folkmann, F., Groeneveld, K. O., Mann, R., Nolte, G., Schumann, S., Spohr, R. 1975: Z. Phys. A **275**, 229

Frischkorn, H. J., Groeneveld, K. O., Schumann, S., Latz, R., Reichhard, G., Schader, J., Kronast, W., Mann, R. 1980: Phys. Lett. A 76, 155

Frischkorn, H. J., Groeneveld, K. O., Schumann, S., Schader, J., Astner, G., Hultberg, S., Lundin, L., Ramanujam, R., Didriksson, R., Hakansson, P., Sundquist, B., Mann, R. 1981: in: *Inner Shell and X-Ray Physics of Atoms and Solids*, Fabian, D. J., Kleinpoppen, H., Watsun, L. M. (Eds.), (Plenum, New York) 193

Frischkorn, H. J., Groeneveld, K. O., Koschar, P., Latz, R., Schader, J. 1982: Phys. Rev. Lett. 49, 1671

Frischkorn, H. J., Groeneveld, K. O. 1983: Physica Scripta T 6, 89

Garnier, H. P., Dumont, P. D., Baudinet-Robinet, Y. 1982: Nucl. Instrum. Meth. 202, 187

Gay, T. J., Berry, H. G. 1979: Phys. Rev. A 19, 952

1979: J. Physique (Paris) 40, C1-298

Goudsmit, S., Saunderson, J. L. 1940: Phys. Rev. 57, 24

Griepenkerl, K., Müller, B., Greiner, W. 1989: Rad. Eff. and Defects in Solids 110, 215

Groeneveld, K. O. 1988: Nucl. Tracks Radiat. Meas. 15, 51

Groeneveld, K. O., Meckbach, W., Sellin, I. A. (Eds.) 1984: *Forward Electron Emission in Ion Collisions* (Springer, Heidelberg) Lecture Notes in Physics LNP213

Groeneveld, K. O., Maier, R., Rothard, H. 1990: Il Nuovo Cimento D 12, 843

Hasselkamp, D. 1985: Habilitationsschrift, Justus-Liebig-Universität, Giessen, Germany

Hasselkamp, D. 1988: Comments At. & Mol. Phys. 21, 241

Hasselkamp, D., Hippler, S., Scharmann, A. 1987: Nucl. Instrum. Meth. B 18, 561

Hasselkamp D. 1991: Springer Tracts in Modern Physics, this volume

Heil, O., Kemmler, J., Kroneberger, K., Köver, A., Szabò, Gy., Gulyás, L., DeSerio, R., Lencinas, S., Keller, N., Hofmann, D., Rothard, H., Berény, D., Groeneveld, K. O. 1988: Z. Phys. D 9, 229

Hippler, S. 1988: Thesis, Justus-Liebig-Universität, Giessen, Germany

Hippler, S., Hasselkamp, D., Scharmann, A. 1988: Nucl. Instrum. Meth. B 34, 518

Hölzl, J., Jacobi, K. 1969: Surf. Sci. 14, 351

Hölzl, J., Schulte, F. K. 1979: in: *Solid State Physics* (Springer, Heidelberg), Springer Tracts in Modern Physics 85, 1

Hofer, W. O. 1989: in: *Fundamental Beam Interactions with Solids for Microscopy, Microanalysis and Microlithography*, Proceedings of the 8th Pfefferkorn Conference, May 7–12, 1989, Park City, Utah, USA; Schou, J., Kruit, P., Newbury, D. E. (Eds.) 1990: Scanning Microscopy Suppl. 4, 265

Holmén G., Svensson, B., Schou, J., Sigmund, P. 1979: Phys. Rev. B 20, 2247

Kemmler, J., Koschar, P., Burkhard, M., Groeneveld, K. O. 1985: Nucl. Instrum. Meth. B 12, 62

Kemmler, J., Heil, O., Biedermann, C., Koschar, P., Rothard, H., Kroneberger, K., Groeneveld, K. O., Sellin, I. A. 1987: J. Physique (Paris) 48, C9-223

Kemmler, J 1988: Dissertation, J. W. Goethe-Universität, Frankfurt am Main, IKF-D-412

Kemmler, J., Heil, O., Biedermann, C., Koschar, P., Rothard, H., Kroneberger, K., Groeneveld, K. O., Köver, A., Szabo, G., Gulyas, L., Berenyi, D., Focke, P., Meckbach W., 1988a: in *High-Energy Ion-Atom Collisions*, Berenyi, D., Hock, G. (Eds.), (Springer, Heidelberg), Lecture Notes in Physics 294, 362

Kemmler, J., Koschar, P., Heil, O., Biedermann, C., Rothard, H., Kroneberger, K., Lencinas, S., Groeneveld, K. O., Sellin, I. A. 1988b: Nucl. Instrum. Meth. B 33, 281

Kemmler, J., Lencinas, S., Koschar, P., Heil, O., Rothard, H., Kroneberger, K., Szabó, Gy, Groeneveld, K. O. 1988c: Nucl. Instrum. Meth. B 33, 317

Kemmler, J., Heil, O., Koschar, P., Biedermann, C., Rothard, H., Kroneberger, K., Meckbach, W., Groeneveld, K. O. 1989a: Z. Phys. D 13, 45

Kemmler, J., Heil, O., Keller, N., Koschar, P., Rothard, H., Kroneberger, K., Sellin, I. A., Groeneveld, K. O. 1989b: Rad. Eff. and Defects in Solids 110, 149

Kemmler, J. 1990: Nucl. Instrum. Meth. B 48, 612

Koschar, P., Clouvas, A., Heil, O., Burkhard, M., Kemmler, J., Groeneveld, K. O. (1987):
  Nucl. Instrum. Meth. B **24/25**, 153
Koschar, P., Szabo, Gy., Clouvas, A., Kemmler, J., Heil, O., Biedermann, C., Rothard,
  H., Kroneberger, K., Lencinas, S., Groeneveld, K. O. (1987b): J. Physique (Paris) **48**,
  C9-275
Koschar, P., Kroneberger, K., Clouvas, A., Burkhard, M., Meckbach, W., Heil, O., Kemmler, J., Rothard, H., Groeneveld, K. O., Schramm, R., Betz, H. D. 1989: Phys. Rev.
  A **40**, 3632
Koyama, A., Shikata, T., Sakairi, H., Yagi, E. 1982: Japanese J. Appl. Phys. **21**, 1216
Koyama, A., Benka, O., Sasa, Y., Ishakawa, H., Uda, M. 1987: Phys. Rev. A **36**, 4535
Koyama, A., Ishakawa, H., Sasa, Y., Benka, O., Uda, M. (1988): Nucl. Instrum. Meth.
  B **33**, 338
Kroneberger, K., Clouvas, A., Schlüssler, G., Koschar, P., Kemmler, J., Rothard, H.,
  Biedermann, C., Heil, O., Burkhard, M., Groeneveld, K. O. 1988: Nucl. Instrum.
  Meth. B **29**, 621
Kroneberger, K., Rothard, H., Burkhard, M., Kemmler, J., Koschar, P., Heil, O., Biedermann, C., Lencinas, S., Keller, N., Lorenzen, P., Hofmann, D., Clouvas, A., Veje, E.,
  Groeneveld, K. O. 1989: J. Physique (Paris) **50**, C2-99
Kroneberger, K., Sigaud, G. M., Albert, A., Heil, O., Maier, R., Rothard, H., Schlösser,
  D., Trabold, H., Groeneveld, K. O. 1991: 14th Intern. Conf. on Atomic Collisions in
  Solids, July 28–Aug. 2, 1991, Salford, UK (Nucl. Instrum. Meth. B, 1992)
Laubert, R., Sellin, I. A., Vane, C. R., Suter, M., Elston, S. B., Alton, G. D., Thoe, R. S.
  1980: Nucl. Instrum. Meth. **170**, 557
Lencinas, S., Burgdörfer, J., Kemmler, J., Heil, O., Kroneberger, K., Keller, N., Rothard,
  H., Groeneveld, K. O. 1990: Phys. Rev. A **41**, 1435
Lorenzen, P., Rothard, H., Kroneberger, K., Kemmler, J., Burkhard, M., Groeneveld,
  K. O. 1989: Nucl. Instrum. Meth. A **282**, 213
  (and Annual Report of the "Institut für Kernphysik der J. W. Goethe-Universität",
  Frankfurt am Main, Germany, IKF-49, 33)
Lotz, W., Burkhard, M., Koschar, P., Kemmler, J., Rothard, H., Biedermann, C., Hofmann, D., Groeneveld, K. O. 1986: Nucl. Instrum. Meth. A **245**, 560
Mazuy, B., Belkacem, A., Chevalier, M., Clouvas, A., Gaillard, M. J., Poizat, J. C., Remillieux, J. 1988: Nucl. Instrum. Meth. B **31**, 382
Meckbach, W., Nemirowsky, I. B., Garibotti, C. R. 1981: Phys. Rev. A **24**, 1793
Meckbach, W., Braunstein, G., Arista, N. 1975: J. Phys. B **8**, L344
Mitschler, J., Benazeth, N., Nègre, M., Benazeth, C. 1984: Surf. Sci. **136**, 532
Musket, R. G. 1975: J. Vac. Sci. Technol. **12**, 444
Neelavathi, V. N., Ritchie, R. H., Brandt, W. 1974: Phys. Rev. Lett. **33**, 302
Oda, N., Nishimura, F., Yamazaki, Y., Tsurubuchi, S. 1980: Nucl. Instrum. Meth. **170**,
  571
Pferdekämper, K. E., Clerc, H. G. 1975: Z. Phys. A **275**, 223
  1977: Z. Phys. A **280**, 155
Rösler, M., Brauer, W. 1988: Phys. Status Solidi (b) **148**, 213
Rothard, H., Burkhard, M., Kemmler, J., Biedermann, C., Kroneberger, K., Koschar, P.,
  Heil, O., Groeneveld, K. O. 1987a: J. Physique (Paris) **48**, C9-211
Rothard, H., Burkhard, M., Biedermann, C., Kemmler, J., Kroneberger, K., Koschar, P.,
  Heil, O., Hofmann, D., Groeneveld, K. O. 1987b: J. Physique (Paris) **48**, C9-215
Rothard, H., Burkhard, M., Biedermann, C., Kemmler, J., Koschar, P., Kroneberger, K.,
  Heil, O., Hofmann, D., Groeneveld, K. O. 1988a: J. Phys. C **21**, 5033
Rothard, H., Lorenzen, P., Keller, N., Heil, O., Hofmann, D., Kemmler, J., Kroneberger,
  K., Lencinas, S., Groeneveld, K. O. 1988b: Phys. Rev. B **38**, 9224
Rothard, H., Kroneberger, K., Burkhard, M., Biedermann, C., Kemmler, J., Heil, O.,
  Groeneveld, K. O. 1989a: J. Physique (Paris) **50**, C2-105

Rothard, H., Kroneberger, K., Burkhard, M., Kemmler, J., Koschar, P., Heil, O., Biedermann, C., Lencinas, S., Keller, N., Lorenzen, P., Hofmann, D., Clouvas, A., Groeneveld, K. O., Veje, E. 1989b: Rad. Eff. and Defects in Solids 109, 281

Rothard, H., Kroneberger, K., Clouvas, A., Veje, E., Lorenzen, P., Keller, N., Kemmler, J., Meckbach, W., Groeneveld, K. O. 1990a: Phys. Rev. A 41, 2521

Rothard, H., Kroneberger, K., Veje, E., Kemmler, J., Koschar, P., Heil, O., Lencinas, S., Keller, N., Lorenzen, P., Hofmann, D., Clouvas, A., Groeneveld, K. O. 1990b: Phys. Rev. B 41, 3959

Rothard, H., Kroneberger, K., Veje, E., Schosnig, M., Lorenzen, P., Keller, N., Kemmler, J., Biedermann, C., Albert, A., Heil, O., Groeneveld, K. O. 1990c: Nucl. Instrum. Meth. B 48, 616

Rothard, H., Schosnig, M., Kroneberger, K., Groeneveld, K. O. 1991a: in *Interaction of Charged Particles with Solids and Surfaces*, Flores, F., Urbassek, H. M., Arista, N., Gras-Marti, A. (Eds.), NATO Advanced Study Institute, Alicante, Spain, May 6–18, 1990 (Plenum, New York)

Rothard, H., Schosnig, M., Schlösser, D., Kroneberger, K., Da Silveira, E., Groeneveld, K. O. 1991b: Nucl. Instrum. Meth. B 56/57, 843

Rothard, H., Schou, J., Groeneveld, K. O. 1991: Phys. Rev. A (1992) in print and Rothard, H., Thesis, J. W. Goethe-Universität, Frankfurt am Main, Germany, 1990)

Sanchez, E. A., de Ferrariis, L. F., Suarez, S. 1989: Nucl. Instrum. Meth. B 43, 29

Sanchez, E. A. 1989: Nucl. Instr. Meth. A 280, 433

Schader, J., Kolb, B., Sevier, K. D., Groeneveld, K. O. 1978: Nucl. Instrum. Meth. 151, 563

Schäfer, W., Stöcker, H., Müller, B., Greiner, W., 1978: Z. Phys. A 288, 349
    1980: Z. Phys. B 36, 319

Schiewitz, G., Biersack, J. P., Schneider, D., Stolterfoth, N., Fink, D., Montemayor, V. J., Skogvall, B. 1990: Phys. Rev. B 41, 6262

Schneider, G. 1931: Ann. d. Phys. 5, 357

Schosnig, M., Rothard, H., Kroneberger, K., Schlösser, D., Groeneveld, K. O. 1991: 2. European Conf. on Accelerators in Applied Research and Technology, Frankfurt am Main, Germany, Sept. 3–7, 1991 (Nucl. Instrum. Meth. B, 1992)

Schramm, R., Koschar, P., Betz, H. D., Burkhard, M., Kemmler, J., Heil, O., Groeneveld, K. O. 1985: J. Phys. B 18, L507

Schou, J. 1980: Phys. Rev. B 22, 2141
    1988: Scanning Microscopy 2, 607

Sellin, I. A., Berry, S. D., Breinig, M., Bottcher, C., Latz, R., Burkhard, M., Folger, H., Frischkorn, H. J., Groeneveld, K. O., Hofmann, D., Koschar, P. 1984: in Groeneveld, Meckbach, Sellin, p. 109

Shi, C. R., Toh, H. S., Lo, D., Livi, R. P., Mendenhall, M. H., Zhang, D. Z., Tombrello, T. A., 1985: Nucl. Instrum. Meth. B 9, 263

Sigmund, P., Tougaard, S. 1981: Springer Series in Chemical Physics 17, 2

Sternglass, E. J. 1957: Phys. Rev. 108, 1

Steuer, M. F., Gemmel, D. S., Kanter, E. P., Johnson, E. A., Zabransky, B. J. 1983: IEEE Trans. Nucl. Sci. NS 30, 1069

Suarez, S., Bernardi, G. C., Focke, P., Meckbach, W. 1988: Nucl. Instrum. Meth. B 33, 326

Thum, F., Hofer, W. O. 1984: Nucl. Instrum. Meth. B 2, 531

Toburen, L. 1989: in *Fundamental Beam Interactions with Solids for Microscopy, Microanalysis and Microlithography*, Proceedings of the 8th Pfefferkorn Conference, May 7–12, 1989, Park City, Utah, USA; Schou, J., Kruit, P., Newbury, D. E., (Eds.)

Toburen, L. 1990: Scanning Microscopy, Suppl. 4, 239

Tougaard, S., Sigmund, P. 1982: Phys. Rev. B 25, 4452

Woods, C. J., Cowern, N. E. B., Bridwell, L. B., Sofield, C. J. 1985: J. Phys. B **18**, 4113
Yamazaki, Y. 1988a: in: *High-Energy Ion-Atom Collisions*, Berenyi, D., Hock, G., (Eds.) (Springer, Heidelberg), Lecture Notes in Physics **294**, 322
Yamazaki, Y. 1988a: Phys. Rev. Lett. **61**, 2913
Yamazaki, Y., Kuroki, K., Komaki, K. I., Andersen, L. H., Horsdal-Pedersen, E., Hvelplund, P., Knudsen, H., Møller, S. P., Uggerhøj, E., Elsener, K. 1990: J. Phys. Soc. of Japan **59**, No 8, 2643
Zaikov, Z. P., Kralinka, E. A., Nikolaev, V. S., Feinberg, Yu. A., Vorobief, N. F. 1986: Nucl. Instrum. Meth. B **17**, 97
Ziegler, J. F., Biersack, J. P., Littmark, U. 1985: *The Stopping and Range of Ions in Matter* (Pergamon, New York)

Slow Particle-Induced Electron Emission from Solid Surfaces*

P. Varga and H. Winter

With 39 Figures

1. Introduction

In the following review we will deal with electron emission from solid surfaces resulting from impact of relatively slow particles (in the given context we denote by *particle* any neutral or charged atom, molecule or cluster, whereas electrons as projectiles remain definitely excluded unless explicitly stated). By *slow* we refer qualitatively to the impact energy range below one keV/amu (impact velocity below 0.2 amu $\approx 4 \times 10^5$ m/s). Phenomena related to particle-induced electron emission ("PIE" – the rather common use of "secondary electron emission" for such processes should be avoided) from solid surfaces can be roughly ascribed to two different mechanisms:

Kinetic Emission ("KE") is caused by transfer of kinetic projectile energy onto the electrons and atomic cores in the solid or in some adsorbates at the solid surface, which may result in a variety of physical processes eventually leading to the ejection of electrons from the solid surface. In KE processes the kinetic energy and mass of the projectile are of foremost importance, whereas a number of other projectile properties as its chemical configuration (atomic or molecular or clustered), charge state (neutral or positively singly or multiply charged or negatively charged) and electronic, vibrational or rotational state (in respective ground state or excited states) are usually not of concern. However, this assumption is only justified at relatively high impact velocity. In the context of the present review we will sometimes check the extent of its applicability. Since KE from its very principle should disappear toward low impact velocity, a corresponding threshold impact velocity should exist, but its precise location is generally not well defined.

The physical processes and experimental experiences related to KE are discussed at length in a review by Hasselkamp (1991) in this volume.

Potential Emission ("PE") is the principal mechanism for electron emission if the kinetic energy of the projectile remains of less influence than its "internal" properties which have just been mentioned. For slow projectiles the charge state plays a rather significant role on the electron emission, and the projectile species is also of concern because of the potential energy related to ion production from the corresponding neutral atomic or molecular ground state species. This potential energy is carried by the ion toward the solid surface

---

*List of Abbreviations see end of this chapter.

and deposited during recombination of projectiles into their neutral ground state. Recombination involves various rapid electronic transitions which may lead to the electron emission.

It is important to note that the PE processes do not require any kinetic projectile energy and therefore no impact velocity threshold is given. However, toward higher impact velocity the PE processes will become superimposed by KE effects which then eventually may dominate (see above).

In the present review we are primarily interested in such low projectile velocities where the PE contributions remain dominant. We will be using the acronym *"sPIE"* for *slow particle-induced electron emission* as a qualitative distinction against KE-dominated processes, which are the subject of other contributions to the present volume.

For most practical applications of PIE the involved impact velocities are not sufficiently low to exclude KE influences completely or even approximately. Therefore it will be necessary to regard the influence of KE in all studies on PE where the projectile velocity reaches up to or surpasses the corresponding KE threshold region.

In the earlier literature on PIE, informations on potential emission are rather scarce and – almost as a rule – not very reliable. Only a few groups have produced still nowadays valid experimental data, mainly because of the very important influence of target surface conditions on the PE process and the comparably small PE contributions in the presence of usually more important KE-related background effects.

In first place we refer to the important studies of H.D. Hagstrum at Bell Laboratories/USA, which started in the early fifties and can be regarded as the foundation of modern PE work both from an experimental (Hagstrum 1953, 1954a, Hagstrum et al. 1965, Chaban et al. 1990) and theoretical (Hagstrum 1954b, 1977) point of view. Seminal contributions to the field of PE have also been provided by experimental work of U.A. Arifov and coworkers in Tashkent/USSR (Arifov et al. 1969, 1971) with the support of a strong theoretical group. Publications from both groups serve as valuable sources for earlier studies in the field. General reviews on both experimental and theoretical contributions from other groups active in the fifties and early sixties have been given by Kaminsky (1965) and Carter and Colligon (1968). Some informations on PE can also be found in treatises on gaseous electronics and related fields, cf. e.g. McDaniel (1964), but to our knowledge there exists no recent extensive survey dealing specifically with PE. However, the processes fundamental to potential emission are commonly included in the numerous surveys on PIE.

Investigations of potential emission have long been plagued by their extreme surface-sensitivity and the related experimental demands on ultrahigh vacuum conditions and target preparation. In fact, Hagstrum has utilized the very surface sensitivity of PE for a particular surface spectroscopy (so-called ion neutralisation spectroscopy – "INS", Hagstrum 1956a, 1978), which later became somewhat superseeded by the closely related metastable atom de-excitation spectrocopy ("MDS", cf. e.g. Ertl 1986).

Although UHV experimental technology has rapidly become widely available, there have not been many important studies on PE in the seventies and early eighties, apart from further contributions by the groups already mentioned.

Still more recently, an important boost to the field has been provided by the growing availability of slow multicharged ions ("MCI") as projectiles. Their use tremendously enlarged the parameter space for PE studies and the importance of PE also at higher impact energy, exposed genuinely new effects and thus led to important further insights in PE-related phenomena just in the last few years. This has also attracted interest from the theoretical side, mainly because of a close relation of PE with the processes of ion-induced desorption, secondary ion emission and low energy ion reflection from surfaces, all of which are of considerable importance for surface investigations. With highly charged projectiles one can also study the phenomenon of "Coulomb explosion" (Bitensky et al. 1979, Bitensky and Parilis 1989) as well as the conversion of approaching highly charged ions into "hollow atoms" which then give rise to characteristic soft X-ray emission (Briand et al. 1990). Both processes should most clearly develop at the lowest achievable MCI impact velocities.

It is therefore quite fitting to speak of a recent renaissance for the field of potential emission as a good justification for the present review which has been organized in the following way.

Chapter 2 will deal with experimental methods for investigation of sPIE and in particular PE. In Chap. 3 a short glimpse on the theoretical background for PE will be followed by discussions of recent experimental data and their interpretation, with particular emphasis on the use of slow multicharged projectile ions and data obtained by the present authors and their affiliates.

Finally, Chap. 4 has been dedicated to the treshold impact velocity region for KE in the presence of PE, where a recently achieved clearer distinction between the two mechanisms will be presented.

Based on this experience we will finally redefine the mechanisms of PE and KE in a physically more stringent way than so far common.

2. Experimental Techniques for Investigation of sPIE

As already mentioned in the introduction, studies on sPIE have been conducted for many decades, but most of the available results should be regarded with reservations because of insufficiently well defined experimental conditions. State-of-the-art investigations in this field must take into account the precise state of the projectile as well as the target surface. As a general rule, projectile particle beams have to be carefully defined according to their geometry and kinetic (impact) energy, and the internal particle energy (possible presence of excited states, etc.) must be known as well (cf. Sect. 2.1).

The target surface has to be prepared by means of now rather common standard techniques, which means that application of UHV technology (base pressure in the target chamber not higher than typically 10^{-10} mbar) is indispensable (Sect. 2.2). In the course of such experiments the reactants (in first place the emitted electrons) should be detected according to their geometry, species and energy (Sect. 2.3). In the present context we deal mainly with neutral and ionized atoms/molecules as projectiles, while electrons are the principal secondary particles. In addition, consideration of processes closely related to PE are of interest (e.g. slow ion scattering, secondary ion emission, ion-induced desorption and photon emission) together with the appropriate experimental techniques.

From the detected reaction products the main features of PE processes can be derived. In first instance the electron emission yield (Sect. 2.4), the energy distribution of ejected electrons (Sect. 2.5) and the electron emission statistics (Sect. 2.6) will constitute a basis for further physical considerations on the processes of interest.

2.1 Preparation of Slow Particle Beams

As the most common projectiles for sPIE investigations we consider singly or multiply charged ions with their impact energy ranging from a few eV up to a few keV. Such ion beams can be produced by a variety of ion sources (Brown 1987), which should be carefully selected in view of the particular needs (ion species, charge states, desired particle fluxes). It is important to recognize that as a general rule the ion source constitutes a most important component of the experimental setup and can greatly influence both the efficiency and reliability of the whole experiment.

An enhanced ion kinetic energy spread caused by the particular ion source may require the incorporation of ion energy/velocity selectors into the beam optics, together with means for mass selection, ion transport and acceleration/deceleration in front of the target surface. Furthermore, the elevated gas pressure resulting from ion source operation might call for a differential pumping stage between the ion source and the target chamber.

Fast (keV range) neutral projectiles can be produced by charge exchange of primary ions in suitable target gases or via ion scattering with neutralization at well defined single crystal surfaces.

Preparation of neutral projectiles with impact energies below typically 100 eV requires special techniques to achieve well defined projectile geometry and kinetic energy (Scoles 1988).

Of particular interest is the characterisation of projectile particles according to their internal energy. The admixture of long-lived excited (metastable) states is often of great influence and should be quantitatively taken into account. A number of techniques have been developed for determination of metastable ion/neutral beam fractions in ion-atom collision experiments (see e.g. Gilbody 1978, Hofer et al. 1983, Schweinzer and Winter 1989) which are directly applicable for the present purposes.

In a number of experiments, fast (multi-keV up to MeV) projectiles have been applied under grazing incidence angles θ in order to obtain small "vertical" impact energies $E_i = E \sin^2 \theta$. However, such a situation is fundamentally different from the case were decelerated ions impinge in normal direction onto the surface, for the following two reasons.

1) The moving particle "sees" the momentum space of surface valence electrons as modified by a Galilei-transformation (van Wunnik et al.1983), which causes an effective broadening of the Fermi edge when seen by the moving projectile, similar to but much more pronounced than thermal broadening. The energetic relations between valence electrons of the surface and the electronic states of an approaching particle are decisive for the electronic transitions to be described in Sect. 3.1. Consequently, electrons with energies above the Fermi edge in the projectile rest frame become available for electron transitions from the surface to the projectile (RN) as well as now empty states below the Fermi edge can serve for transitions from the particle to the surface (RI). This Galilei transformation also causes an increase in energy of electrons emitted from the valence band with increasing parallel ion velocity (i.e. a similar effect as a decreasing work function, Mišković and Janev 1989) and also permits processes to take place which would be energetically forbidden for ions with low impact energy.

2) Precautions are necessary to avoid violent atomic collisions with single surface atoms which would dominate any effects due to the potential energy of the projectile. First of all, targets with a perfect flat surface (i.e. flat on an atomic scale, which means relatively large, well oriented and prepared single crystals with extremely few terraces) have to be used. Secondly, as has been pointed out recently by Winter and Zimny (1988), for grazing incidence and a planar scattering potential the scattering parameter has to be chosen such that the distance of closest approach remains larger than the so called "Firsov screening length"

$$a_{\rm F} = 0.885 a_0 \left(Z_1^{1/2} + Z_2^{1/2} \right)^{-2/3} \qquad (2.1)$$

with a_0 the Bohr radius and Z_1, Z_2 the projectile and target atomic numbers, respectively. The critical angle θ_c for this condition to prevail can be calculated from the characteristic planar channeling angle (Gemmell 1974).

$$\theta_c = \sqrt{\frac{2\pi Z_1 Z_2 e^2 a_{\rm F} n_s}{E_0}} \qquad (2.2)$$

153

with n_s the planar density of atoms at the surface and E_0 the primary ion energy.

Important contributions from violent collisions causing kinetic emission processes have to be expected in all experimental situations where these rather stringent limitations for grazing incidence experiments are not met.

2.2 Target Preparation

Electron emission from low energy ion impact results almost exclusively from interactions of the projectile with the very first surface layer. Therefore, the electronic density of states of the surface together with the electronic state of the ion determines the emission yield, energy distribution, angular dependence, etc. Only carefully prepared surfaces will assure reproducible data for PE investigations.

In appropriate experiments the usual requirements in surface physics have to be carefully fulfilled, and suitable in situ-surface analytical techniques must be incorporated. This is the reason why data on electron emission which have been performed in vacua above typically 10^{-9} mbar are only of limited significance. It is important to know the chemical composition of the surface with respect to adsorbates, since adsorbed sub-monolayers can alter the work function, the density of states of the valence band and thus the electron emission, as well.

In particular, for single crystal surfaces the crystallographic orientation is of interest, especially if angle-dependent investigations are performed and low impact energy is simulated by grazing incidence of fast particles.

There are two different possibilities to assure reproducible data.

1) Use targets for which many data on preparation techniques are available and which are known to pose no great "difficulties" for preparation.
2) Still more appropriately, in situ surface-sensitive analytical methods like AES (Auger-Electron Spectroscopy), PES (Photoelectron Spectroscopy) as well as ISS (Ion Surface Scattering) should be used.

A clean surface may be defined as to be annealed and at an ambient temperature with an overall surface contamination of less than a few percent of a monolayer (10^{13} atoms /cm^2). At any case, appropriate vacuum conditions are necessary to keep a such prepared surface clean as long as the measurements takes place.

The preparation of surfaces is extremely material-dependent, but in general the possibilities for noble gas ion sputtering (see the recent review by Taglauer 1990), heating of the surface and flooding with reactive gases are necessary for target preparation. In most cases heating in good vacuum causes impurities of the bulk material to segregate (enrichment) at the surface which must than be sputtered away or removed by heating under elevated oxygen or hydrogen gas pressure. By repetitive application of these methods (sometimes necessary during several days) a clean and reproducible surface can be prepared.

A useful summary of important techniques for the preparation of clean surfaces after a sample has been brought into an ultra-high vacuum environment (i.e. after mechanical and/or electro-chemical polishing and of residue removal) can be found in Musket et al. (1982) and Grunze et al. (1988).

For a number of elements the only way to achieve a clean surface for the present purposes is either in-situ deposition of thin films via sputtering or evaporation, or in-situ fracturing or cleaving of appropriate samples.

2.3 Detection of Reaction Products

Investigation of particle-induced electron emission in first instance involves quantitative detection of the electrons emitted from the target surface.

In the most simple situation this can be achieved by measuring the current of these electrons, given sufficiently high projectile particle fluxes and emission yields (electron currents at least in the pA range). However, in the majority of experimental situations the projectile fluxes are too low for such an approach and thus the resulting small electron fluxes can only be detected by means of counting techniques (application of open multi-stage electron multipliers, channel electron multipliers, channel plates, etc.). This is especially relevant for the angular dependence or the energy distribution of emitted electrons. Appropriate measures have to be taken to assure constant electron detection efficiency irrespective of the electron impact energy and incidence angle at the active detector region, and also to avoid disturbing effects from heavy particle impact (reflected neutral and/or ionized projectiles, secondary ions) as well as energetic photons "seen" by the active detector region. Special problems can also arise in view to the compatibility of the electron detector with the necessary UHV preparation procedures (baking of UHV apparatus, target cleaning etc., cf. Sect. 2.2).

2.4 Measurement of Electron Emission Yields

Electron emission yields (i.e. the mean number of electrons emitted per impinging projectile particle, usually denoted by γ (see also Sect. 2.6) are determined by measuring the fluxes of both the projectile particles and the emitted electrons (cf. Sect. 2.3). With charged projectiles this can be simplified by measuring target currents with and without permitting the electrons to leave the target, which can be controlled by an appropriate target bias with respect to its environment. Precautions have to be taken against possible disturbances from charged particle reflection, secondary ion emission and spurious electron production due to impact of reflected or scattered projectiles or electrons, all possibly causing additional electron emission from the target region. In general, for such measurements projectile fluxes corresponding to not less than one nA are necessary. For smaller fluxes the electron yield can only be measured by counting both the projectiles and the emitted electrons and then calibrating such obtained relative figures to already known absolute yield data. In such experimental situations the usually rather limited acceptance angle of the electron detector calls for a series of measurements at different electron ejection angles, from which the correct values for the global electron yield have to be derived.

This is especially important because the often made assumption of simple dependences for the electron emission yield on the ejection angle can introduce large experimental errors.

The use of neutral projectiles poses special problems, because determination of their flux requires relatively complicated experimental techniques. Whereas with sufficiently high projectile fluxes microcalorimetric techniques can be applied, recently the use of liquid He-cooled semiconductor bolometers for detection of rather small neutral particle fluxes has become common (cf. Scoles 1988). The latter technique can also be utilized for the present purposes. However, incorporation of suitable devices into the common experimental setups for PIE investigations requires considerable constructional efforts, in particular for the liquid He cryostat needed for cooling of the bolometer unit.

2.5 Measurement of Ejected Electron Energy Distributions

Setups for the joint measurement of electron emission yields (cf. Sect. 2.4) and the energy distribution of emitted electrons in connection with slow particle impact on solid surfaces are usually derived from Hagstrum's instruments (Hagstrum 1954a, Chaban et al. 1990). They may include up to three highly transparent grids inside a hemispherical collector (cf. Hofer et al. 1983), in a similar way as for conventional low energy electron diffraction (LEED) systems. Measurements with this type of equipment will accept electrons with practically zero ejection energy and their emission within almost half of the solid angle, but are only applicable with total electron currents not smaller than typically ten to hundred pA. Moreover, they feature only rather poor electron energy resolution (≥ 1 eV FWHM, cf. Hasselkamp 1991).

As such comparably high electron currents are not commonly available, various standard types of electron energy analysers equipped with electron counting detectors are being applied (see e.g. de Zwart et al.1989, Zeijlmans van Emmichoven et al. 1988), which permit considerably higher energy resolution (≤ 0.1 eV FWHM), but may seriously discriminate the low energy part of the electron energy distribution $K(E_e)$. They also cover rather small solid angles of ejection, which makes a series of measurements at different ejection angles necessary.

2.6 Measurement of Electron Emission Statistics

In sPIE processes involving neutral and singly charged particles the observed yields are typically smaller than unity. If KE is the exclusive (i.e. for neutral projectiles) or dominating process, the electron emission will exhibit a statistical behaviour, i.e. for impact of a single projectile particle ejection of either zero or one, two, etc. electrons from the target surface can result. Consequently, the electron emission yield γ is actually the weighted mean of the so-called electron emission statistics (ES), which itself is represented by the series of probabilities W_n for the emission of a given number n of electrons:

$$\gamma = \sum_{n=1}^{\infty} n W_n \qquad \text{with} \qquad \sum_{n=0}^{\infty} W_n = 1 \quad . \tag{2.3}$$

The ES for PIE have been of longstanding interest in the context of single particle counting with ion-electron converters and secondary electron emission multipliers, because the corresponding counting losses are determined by the probability W_0 for emission of no electron. Studies of ES have therefore been carried out by various groups for a large number of projectile-target combinations (e.g. Bernhard et al. 1965, Delaney and Walton 1966, Staudenmaier et al. 1976).

In these studies the primary emphasis was to clarify precise shapes of the emission probability distributions, since fitting of an appropriate standard distribution (e.g. a Poissonian) to the experimentally determined relative values for W_n with $n \geq 1$ could lead via extrapolation to the not directly observable value for W_0 and thus to the absolute electron emission yields.

Recently, such ES measurements have been extended to much lower ion impact energies than so far attainable (Lakits et al. 1989a, 1990a, Aumayr et al. 1991). As a consequence, comparison of sPIE data resulting from bombardment of a given target surface with isoenergetic neutral, singly and multiply charged projectile species has become possible. This opened up the way for a clearer distinction between contributions from respectively PE and KE to apparent total electron emission data (for further details cf. Sect. 3.4 and Chap. 4).

3. Review on Potential Electron Emission Processes

In the present chapter we will deal exclusively with electron emission caused by the *potential energy* of a projectile particle that approaches toward (or recedes from) a solid surface.

As already stated at several occasions, only particles moving with relatively low kinetic energy with respect to the surface will cause exclusively potential emission, as a result of their internal energy (i.e. because of their being ionized and/or highly excited), whereas at higher impact energies the total (or apparent) electron emission can also be caused by kinetic processes. Contributions from PE and KE are generally not separable in an unambiguous manner. However, from a measurement of electron statistics for PIE processes such a distinction among PE and KE effects has become possible, as will be further discussed in Chap. 4.

In the context of Chap. 3, possible KE contributions will not be further dealt with except as a background for the PE processes of interest.

3.1 Models for Potential Electron Emission

Potential emission (PE) is caused by neutralization and deexcitation processes taking place in front of the surface, where essentially no kinetic energy of the projectile is needed to induce these processes. In the following we will deal with phenomena of electron emission due to deexcitation and neutralization of excited atoms as well as singly and multiply charged ions (MCI) at solid surfaces.

The standard model for electron emission resulting from the interaction of slow ionized or excited particles with surfaces is due to Hagstrum (1954b), who developed a semiempirical theory on the basis of his own experimental results and earlier considerations, e.g. from Shekter (1937).

In this model he regarded charge exchange processes between the electronic states of the surface and the incident particle. These electronic transitions were described within an adiabatic picture which requires no exchange of energy between electronic and nuclear motion. The involved energy levels are slightly shifted upwards, thus leading to a lower binding energy than for the situation of a free particle far from the surface, which primarily results from the interaction of the active electron with the negative image charge of the positively charged projectile core.

Hagstrum obtained transition rates for electronic transfer from the surface valence band to a particle in front of this surface, which depended roughly exponentially from the surface-particle distance d. The transition rates are related to the overlap of the atomic wave functions of surface atoms and the projectile, which both in first approximation decay exponentially with d. All quantum-mechanical calculations of transition rates which have been carried out later on (e.g. Hentschke et al. 1986) showed the validity of this assumption. Consequently, the highest transition probability is to be expected between surface states with the lowest binding energy (i.e. those situated near the Fermi edge) and highly excited projectile states.

Hagstrum also regarded the transition probability as independent from the ion impact energy. He showed that the distance where these transitions take place most probably (i.e. for the maximum of the transition probability) is about 3 Å for He^+ with 10 eV kinetic energy, and moves closer to the surface with increasing particle velocity.

In a rough approximation, the distance d_q where multiply charged ions start to interact with a surface has been shown by Delaunay et al. (1987a) to be proportional to the charge state q.

In a more detailed approach, Apell (1987) assumed that these transitions start at a distance d_q where the ionic Coulomb potential becomes equal to the binding energy of electrons at the Fermi edge, and for a q-times ionized atom he derived a distance of

$$d_q\,[\text{Å}] \approx 3q \ . \tag{3.1}$$

With a similar "over barrier model" and additional inclusion of the electron and ion image potentials Snowdon (1988) obtained

$$d_q\,[\text{Å}] \approx q + 3.7 \ , \tag{3.2}$$

which for highly charged ions becomes much smaller than the value given by (3.1).

A charged particle in front of a metallic surface causes the metal electrons to form image charges. The image plane does not coincide with the surface of the top layer of atoms, because the metal electrons will penetrate by about 1 Å into vacuum .

In a general view, the ion-surface system is subject to two influences (see Varga 1987 and Andrä 1989).

1) The ion "feels" an attractive potential and thus will be accelerated toward the surface, gaining kinetic energy by the classical image force until it has become fully neutralized. A simple calculation (Varga 1987) of this energy gain ΔE_{kin} normal to the surface with the earlier derived distance of interaction (3.1) yields

$$2.5\,q \le \Delta E_{\text{kin}}\,[\text{eV}] \le 4q \ . \tag{3.3}$$

2) The potential energy of the ion-surface system changes with the mutual distance d, where for $d \le 2\,\text{Å}$ the classical image potential concept remains applicable. At smaller distances the potential energy is usually assumed to remain constant, simply to avoid divergence problems. The valence electron to be captured by the incoming ion "sees" besides the field of the ion core also the field of its own image charge and the image charge of the ion core, which for positive ions yields an eigenstate which is shifted upwards, i.e. to lower binding energies. For sufficiently large distances ($d \ge 2\,\text{Å}$) from the surface the resulting binding energy shift $\Delta W = W_i - W_i'$ can be described by

$$\begin{aligned} \Delta W &\approx (2q - 1)/4d \\ \Delta W\,[eV] &\approx 3.6\,(2q - 1)/d[\text{Å}] \end{aligned} \tag{3.4}$$

with W_i and W_i' the potential energies for infinite separation and a distance d, respectively.

The linear relationship between the initial charge state and the corresponding distance of first interaction (with $d[\text{Å}] \approx 3q$ from (3.1), (3.4) yields $\Delta W \, [\text{eV}] \approx 2.4 - 1/q$) leads to the estimate, that for each neutralization step (i.e. decrease of q by one) the binding energy is shifted in first approximation by $\Delta W \geq 1.2$ eV for low charge states, which increases to $\Delta W \leq 2.4$ eV with a higher initial charge state. This shift reduces the potential energy available for electron emission. Since the energetic relations between electrons of the surface and the approaching particle are decisive for the processes to be described below, effects connected with the projectile velocity have to be considered, especially if low "vertical" velocities are simulated by grazing incident projectiles (cf. Sect. 2.1)

The Galilei-transformed broadening of the Fermi edge in the frame of the moving projectile (van Wunnik 1983) in the model presented in Fig. 3.1 makes electrons in states above the Fermi edge available for RN processes as well as opens up states below the Fermi edge for RI transitions. The same Galilei transformation also causes an increase of the kinetic energy of electrons emitted from the valence band with increasing parallel ion velocity in the same way as a decreased work function (Mišković and Janev 1989).

It should be pointed out that this energy shift depends on the total ion velocity and not just on the surface-normal velocity component, the latter however being responsible for the amount of image charge-induced energy shift of the binding energy in the moment of the electron transition.

In Fig. 3.1 electron binding energy schemes for three different classes of transitions are shown. We distinguish one-electron transitions as resonance ionization (RI), resonance neutralization (RN) and quasiresonance neutralization (QRN), from two-electron transitions as Auger neutralization (AN), Auger deexcitation (AD) and autoionization (AI). Finally, also radiative deexcitation (RD) might be possible. W_i' and W_{ex}' are the ionization and excitation energy of the particle near the surface, respectively. The time scale for transitions of interest covers the range of $10^{-13} - 10^{-16}$ s if not otherwise specified.

In the following these processes will be discussed in somewhat more detail.

One-electron Transition Processes

Resonance transitions. Resonance transitions into excited projectile states (cf. Fig. 3.1a) generally dominate the particle-solid interaction because of the much larger spatial extension of the involved wave functions in comparison to the corresponding ground states. As a consequence, such resonance transitions can take place already at a rather large distance from the surface.

Resonance processes are non electron-emitting precursors for the subsequently possible electron-emitting two-electron transitions.

Resonance Neutralization (RN). RN involves electron transfer from the surface valence band into the incident ion and can take place if unoccupied electronic states of the atom become energetically degenerate with those of

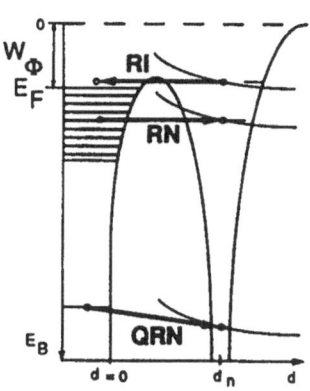

Fig. 3.1a. Resonance transitions

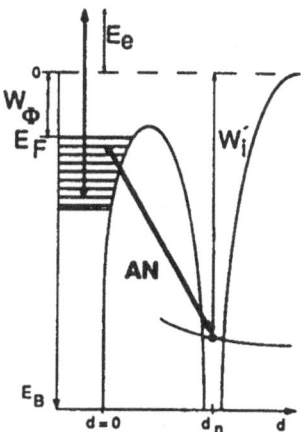

Fig. 3.1b. Auger neutralization

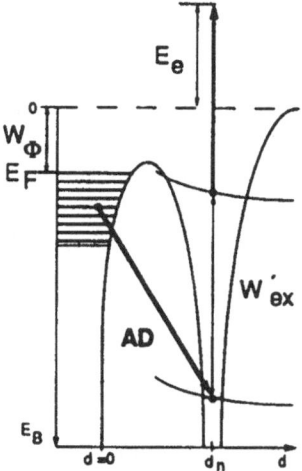

Fig. 3.1c. Auger deexcitation

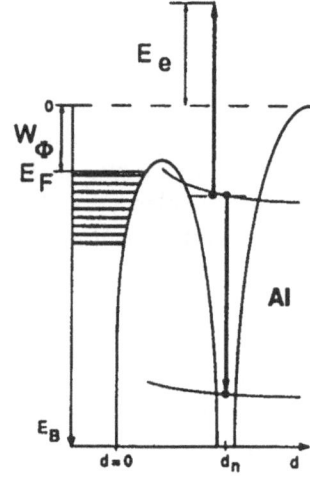

Fig. 3.1d. Autoionization

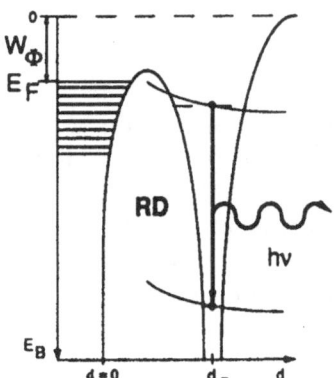

Fig. 3.1e. Radiative deexcitation

Fig. 3.1 a-e Electron energy diagrams showing transition- and deexcitation processes for an atom in front of a surface, E_B ... binding energy, d ... particle-surface distance. The shaded region is the filled portion of the conduction band of the solid surface, W_ϕ ... work function, E_F... Fermi energy. Full dots indicate occupied states, open circles indicate free states. The shift of the electron energy levels with distance is indicated qualitatively and the respective binding energies are denoted with W_i' or W_{ex}'. E_e ..... kinetic energy of possibly emitted electrons

electrons in the valence band. In a potential curve diagram such RN transitions correspond to crossings between the initial (ion plus solid) and final state (excited neutralized particle plus solid).

Resonance Ionization (RI). Resonance ionization (RI) is inverse to RN and can take place if the binding energy of an occupied excited state in the particle is smaller than the surface work function W_ϕ, i.e. if empty levels in the conduction band become energetically resonant with occupied atomic levels.

Quasi-resonance Neutralization (QRN). QRN is a near-resonant transition from more tightly bound localized target states (core states) to projectile states. Such processes can only occur in close collisions since an interpenetration of the involved inner electronic orbitals has to be assured. QRN was found responsible for oscillatory ion survival probabilities in ion-surface-scattering (ISS) experiments (Erickson and Smith 1975).

Two-electron processes

Auger neutralization (AN). This process (cf. Fig. 3.1b), by decreasing the charge state of the ion, can eject an electron from the surface valence band if the involved neutralization energy is at least twice the work function W_ϕ. Two electrons of the surface valence band will be involved, one neutralizing the ion and the other one, gaining energy via electron-electron interaction, to be ejected with a maximum kinetic energy E_e

$$E_e \leq W_i' - 2W_\phi \tag{3.5}$$

W_i' is the effective recombination energy of the neutralized particle ("neutralization energy"), which decreases with the AN process taking place closer to the surface because of an increasing level shift with decreasing distance. The energy distribution of electrons emitted due to AN corresponds to a self-convolution of the electronic surface density of states (S-DOS) and depends on the Auger transition matrix elements and the available potential energy W_i'. An increasing impact energy shifts the peak of the ejected electron distribution to lower kinetic energy, while its maximum energy will be increased because of the collisional broadening of the distribution.

Auger deexcitation (AD). An excited particle (cf. Fig. 3.1c) in the absence of available empty resonant levels in the valence band (i.e. if the binding energy $(W_i' - W_{ex}')$ of the excited electron in the particle is larger than the surface work function W_ϕ) may interact with an electron of the valence band. The such excited electron can be ejected from the particle while a surface electron is transferred into a lower particle state. The maximum energy of a such emitted electron is given by

$$E_e \leq W_{ex}' - W_\phi \tag{3.6}$$

with W'_{ex} (excitation energy) being independent of the particle-surface distance and therefore also of the particle velocity, since the initial and final charge states are identical and thus subject to similar level shifts. The energy distribution of electrons emitted due to such an AD process directly reflects the electronic surface density of states (S-DOS) multiplied by an energy-dependent escape probability. In first order no shift of the related electron energy distribution is to be expected for changing the particle impact energy.

Autoionization (AI). This process can be ascribed to an intra-atomic Auger deexcitation of a doubly/multiply excited atom or ion, with one or more electrons being emitted and one transferred to a lower state of the particle (cf. Fig. 3.1d). AI is of particular interest for MCI neutralization, during which multiple excited states with high quantum numbers might be produced via multiple RN. AI is similar to the AD process, with all involved electrons belonging however to the projectile atom. Therefore, a comparably narrow energy distribution of emitted electrons can be expected. The energy E_e of emitted electrons as in the other cases results from the potential energy difference of the particle before and after the transition. Since AI always increases the charge state at least by one, the energy E_e will be increased by a value of ΔE_e, if particles decay near a surface as compared to AI in free space, because the level shift ΔW for the final state (charge state $q + 1$) is larger than for the initial state (charge state q). ΔE_e increases with the transition occuring closer to the surface, because the difference in the level shifts increases with decreasing distance (Hagstrum and Becker 1973). From (3.4) we can estimate the value of ΔE_e by

$$\Delta E_e \, [\text{eV}] \approx 1.8/d \, [\text{Å}] \; . \tag{3.7}$$

Radiative Deexcitation (RD). For singly charged ions approaching a surface, getting rid of the deexcitation energy by photons (cf. Fig. 3.1e) is highly improbable as compared to the above described radiationless electron transitions, because the involved radiative lifetimes of typically 10^{-8} s is about 10^6 times longer than the transition times for radiationless deexcitation. However, for highly charged ions radiative deexcitation can become competitive because of their rapidly increasing radiative transition rates (for hydrogen-like ions the radiative transition rate is proportional to the fourth power of the charge state, Bethe and Salpeter 1957). Donets (1983) and Donets et al. (1985) have observed X-rays for impact of slow ($E_i \approx 0.4 \, \text{keV/amu}$) Ar^{17+} and Kr^{35+} ions (both ions being hydrogen-like) on Be and Cu targets, respectively. Observed emission of photons clearly corresponded to $1s - 2p$ and $1s - 3p$ transitions, and even some contributions from $1s - np$ (with $n \geq 4$) transitions in multiply excited Ar and Kr atoms have been observed. The photon energy corresponding to the $1s - 2p$ transition was shifted toward higher energy with respect to the well known K_α line from a free atom, but still at somewhat lower energy as compared with the estimated $1s - 2p$ transition in He-like Ar or Kr ions. This result clearly shows that in the most probable case a $2p$ electron jumps to the $1s$ state when the MCI is almost completely neutralized.

Recently, X-ray emission due to impact of Ar^{17+} with 340 keV on Ag has been studied (Briand et al. 1990). In this paper the conclusion had been obtained that by the multiple electron capture processes into high projectile levels so called "hollow" particles are being formed. These particles contain inner shell vacancies, which become later filled via emission of X-rays.

Calculation of PE Transition Rates. Only a few papers have been published with quantitative considerations of the yield and also the energy distribution for PE emission (Hagstrum 1954b, Hofer 1983, Hofer and Varga 1984, Wouters et al.1989, Zeijlmans van Emmichoven et al. 1990). All these calculations have been performed by modeling the emitted energy distributions with the help of semiclassical approximations. From a known S-DOS (Hofer 1983, Hofer and Varga 1984, Wouters et al. 1989 and Zeijlmans van Emmichoven et al. 1990) or assuming a constant S-DOS (Hagstrum 1954b) the electron energy distribution can be simulated in a first approximation if all possible transition processes like AN, AD or AI are regarded.

Much more efforts have been devoted to calculate transition rates for the above mentioned neutralization- and deexcitation processes. The driving force behind this work has been the quantification problem in surface analytical techniques like ISS (ion surface scattering spectroscopy) and SIMS (secondary ion mass spectroscopy). Therefore and because of practical reasons (see below) mainly transition rates for resonant charge transfer (RI and RN) have been investigated (see a recent review by Amos et al. 1989).

In all quantum mechanical calculations of particle surface interaction it is assumed that the motion of the atomic nucleus can be treated classically, thus involving a classical trajectory $R(t)$. The total Hamiltonian for the system can then be reduced to one for the electronic motion only, associated with a one-electronic Hamiltonian $\mathcal{H}(R) = \mathcal{H}(t)$ which depends parametrically on the nuclear position and via the latter on time.

Therefore, the problem becomes one of solving a time-dependent Schrödinger equation

$$\mathcal{H}(t)|\phi(t)\rangle = i\partial|\phi(t)\rangle/\partial t \tag{3.8}$$

for the electronic wave functions $|\phi(t)\rangle$ representing the particle and the valence band of the surface, respectively.

It is hardly possible to obtain an exact form for this Hamiltonian. Progress has been made by Blandin et al. (1976) and Bloss and Hone (1978) by adopting the time-dependent Anderson-Newns (TDAN) Hamiltonian which is a generalization of the time-independent one originally introduced by Anderson (1961) and Newns (1969). It was found that the Anderson-Newns Hamiltonian contains most of the physically important aspects of the problem. Even though this model leads to formally simple equations, it appears to be impossible to obtain analytical solutions. In order to extract physically meaningful results it is necessary to resort to numerical methods or to apply approximations. One of them is to describe resonant charge transfer (Amos et al. 1989) within the frame work of one-electron theory, i.e. by approximating the Hamiltonian by the sum of one-electron operators. This theory has been quite successful in explaining much of the experimental data.

Since Auger transitions play an important role in many particle-surface scattering systems, a more rigorous theoretical treatment is required as has been shown e.g. by Hentschke et al (1986) and Snowdon et al. (1986), who calculated transition rates for AN of low energetic protons scattered on Au and Al. They found that the transition rate over a wide range of ion-surface separations d can be well represented by the simple exponential form Ae^{-ad}, where A and a are constants. This dependence has already been adopted by Hagstrum (1954b).

The conclusion that the theory of surface-charge transfer should be formulated by using many-electron wavefunctions was already drawn by Tully (1977) who laid the foundation for this theory. Only a few many-electron calculations have been made so far, as for example by Kasai and Okiji (1987), Brako and Newns (1985a,b), and Sulston et al. (1988). The results differ qualitatively from those of the one-electron treatments. At the moment most of the calculations are restricted to surfaces which are represented by a noninteracting Fermi gas with a realistic work function and particles with a single, non-degenerated valence level with a realistic binding energy.

A major theoretical effort should now be devoted not only to obtain more quantitative results but also to develop a framework which can encompass all types of charge-transfer processes, including not only the resonance-, but also Auger- and quasi-resonant transitions. The application to more realistic systems should than be a next step.

Recently, Bardsley and Penetrante (1991) have developed an interesting classical approach to treat PE for impact of highly charged ions on metal surfaces. They regarded the PE process as a tunneling of electrons from the metal valence band via the distance-dependent potential barrier into vacuum, under the combined influences of the (point like) projectile charge and its image charge. Electron motion is assumed as classical and a Monte Carlo method is applied to calculate yields and energies of the electrons extracted into vacuum. The calculations are stopped at a projectile distance of several Ångstroms in front of the surface, where quantum effects on the final electron energies can no more be neglected. These calculations provide interesting new aspects for the evolution of PE during the approach of highly charged slow ions toward a metal surface.

3.2 Experimental Results on PE – Total Electron Yields

The yield (or more precisely total yield) γ of electrons emitted from a target surface as a result of PE is defined as their total number per impinging projectile particle

$$\gamma = N_e/N_p = qI_e/I_q \tag{3.9}$$

where N_e and N_p are the fluxes (particles per second) of electrons emitted from and projectiles arriving at the target surface, and I_e and I_q are the corresponding electric currents of electrons and initially q-times charged projectile ions, respectively.

At least for single crystal surfaces strong deviations from an often assumed smooth take-off angle dependence of the angular-differential potential emission yield might occur. However, in the context of this review we will not discuss this feature in more detail, since systematical measurements for both single-crystalline and polycrystalline surfaces are lacking.

An impressive amount of PE total yield measurements can be found in the older literature (cf. references in Kaminsky 1965, Carter and Colligon 1968). Unfortunately, the greater part of these data cannot be regarded as definite because of reasons already explained in Chap. 1.

The most extensive and reliable work is due to H.D. Hagstrum (see a number of corresponding references in Kaminsky 1965, Carter and Colligon 1968), who has thoroughly investigated the PE process in dependence on charge state and species of projectiles as well as the material and surface conditions of targets materials. Most of Hagstrums investigations have been made with singly charged noble gas ions in their ground state, but long-lived highly excited ("metastable") atoms, singly charged molecular and metastable highly excited atomic ions as well as multicharged ions have also been applied for a number of investigations. Furthermore, measurements have been carried out for different polycrystalline target materials in both atomically clean state and with varied coverage of some adsorbates as H_2, N_2, CO_2 etc. The typical projectile impact energy range was chosen from a few eV up to 1 keV.

A large amount of yield measurements, primarily with singly charged ions as projectiles, has been performed by Arifov et al. (1969), including investigations with single crystal targets for metals as well as insulators.

In the following, by way of examples from our own investigations, we will shortly review the typical dependence of the total PE yield on the potential energy and impact (i.e. kinetic) energy of projectile ions, and on the target surface conditions. We will not discuss the influence of the projectile impact angle, for which the reader is referred to Sect. 3.1.

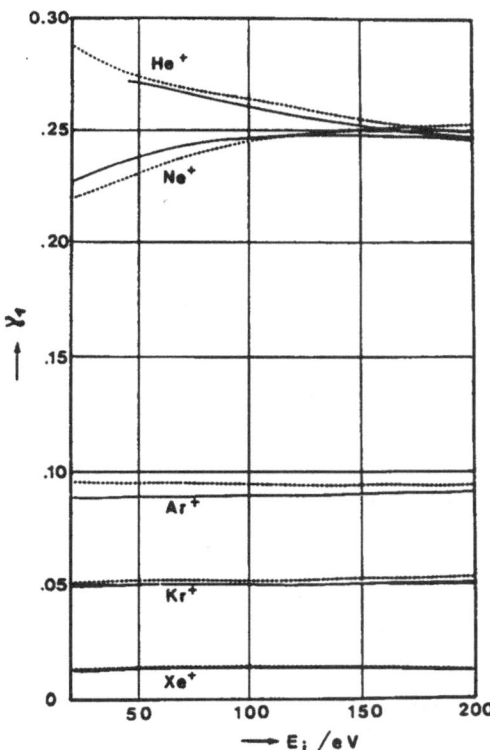

Fig. 3.2. PE yield γ for impact of singly charged noble gas ions on clean polycrystalline tungsten at kinetic energies between 20 and 200 eV (full line, Varga 1978). Dotted line represents results taken from Hagstrum (1956b) for comparison

3.2.1 Singly Charged Projectiles

Figure 3.2 (adapted from Varga 1978) shows results obtained for perpendicular impact of singly charged ground state noble gas ions on clean polycrystalline tungsten in comparison with equivalent results from Hagstrum (1956b). The latter data originated from earlier ones (Hagstrum 1954a) after their correction for the influence of metastable admixtures in the projectile fluxes (see also below). Note the very good agreement despite a time span of more than twenty years between both sets of investigations, the very pronounced dependence on the projectile species (i.e. the involved potential energies) and the relatively weak influence of the projectile impact energy in the range covered, except for the Ne^+ projectiles, where a marked decrease of γ can be observed in the lowest impact energy range. This has been explained by Hagstrum (1954b) to result from the contribution of RN-AD to the neutralization of Ne^+ on clean W, whereas for all other ion species only AN processes can take place.

From the bulk of available data, which have been obtained with a variety of projectile ions impinging on both clean and gas-covered target materials, a qualitative dependence of γ on the projectile potential energy W_i and the target surface work function W_ϕ can be formulated

$$\gamma = \alpha \left(\beta W_i - 2 W_\phi \right) \quad . \tag{3.10}$$

Note, that this relation does not contain the projectile impact energy, cf. above. The shape of (3.10) has been theoretically derived by Kishinevskii (1973), who obtained $\alpha = 0.2/\varepsilon_F$ with ε_F the target surface Fermi energy, and $\beta = 0.8$. Using a least square fit to available experimental data, Baragiola (1979) obtained $\alpha = 0.032$ and $\beta = 0.78$ for a number of target species (all energies given in eV).

According to this relation, γ will apparently increase with the projectile potential energy and with a decreasing surface work function. This is also clearly demonstrated by γ data obtained with respectively clean and electronegative gas-covered target surfaces, cf. Hagstrum (1956d) and Fig. 3.3 from Hofer (1983). Increase of the oxygen coverage corresponds to an increasing work function. In the same figure also the influence of W_i is nicely demonstrated by comparing data for respectively ground state and metastable projectiles, as has first been demonstrated by Hagstrum (1956b, 1960) and later confirmed by Varga and Winter (1978), cf. Fig. 3.4.

Available results with molecular ions follow same tendencies as observed with atomic ions.

Apart from the influence of the surface work function, no clear relation between the S-DOS (surface density-of-states) of a target and the corresponding PE yield has been demonstrated, whereas the ejected electron energy distribution can be strongly influenced. The latter fact has also been utilized for the development of a particular surface spectroscopy ("INS", cf. Sect. 3.3).

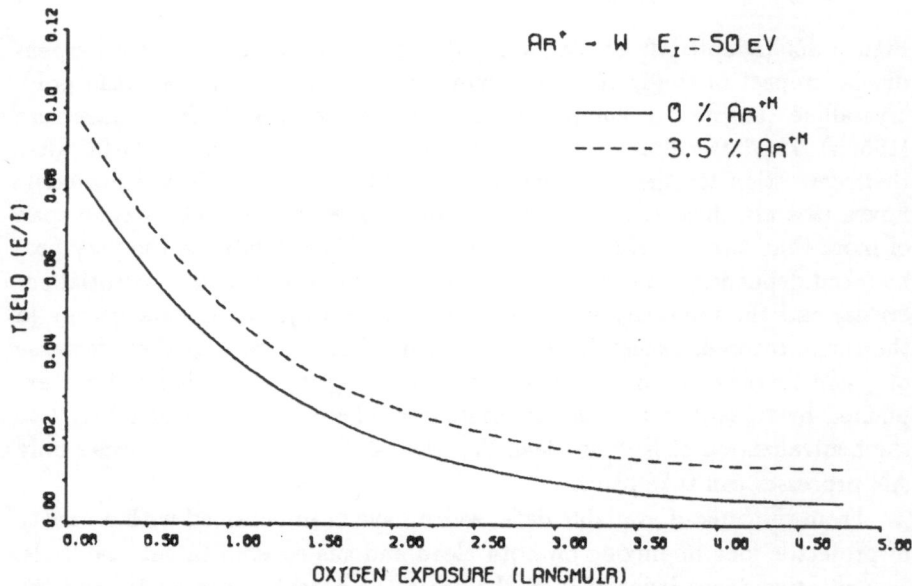

Fig. 3.3. PE yield γ induced by impact of 50 eV singly charged Ar^+ ions on tungsten without (full line) and with 3.5 % admixture of highly excited metastables (dashed line), for varying oxygen coverage (Hofer 1983)

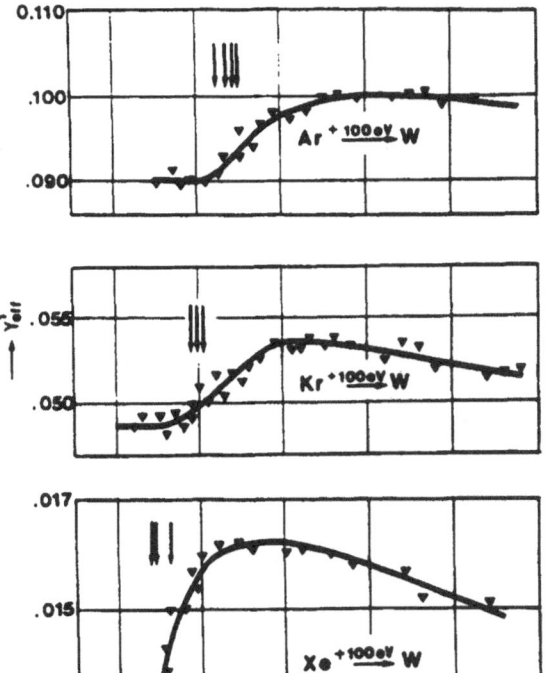

Fig. 3.4. Effective PE yield γ from clean tungsten for impact of 100 eV singly charged noble gas ion beams with metastable ion admixture dependent on E_{IQ}, the electron impact energy in the ion source. Appearance potentials for metastable states have been indicated by arrows (Varga and Winter 1978)

3.2.2 Excited Neutral Projectiles (Metastable Atoms)

Several groups have investigated the long-known fact that highly excited slow atoms can give rise to PE (cf. references in Kaminsky 1965). This emission is usually caused by AD occuring in front of the surface, from where electrons will be ejected if the available excitation energy exceeds the surface work function. For many metastable atoms their excited states energetically coincide with states populated via RN from the surface into a singly charged ground state ion of the same species. In such cases the metastable atom yields should be about the same as for the corresponding ground state ions. For a further discussion of this situation cf. Sect. 3.3.1. AD of metastable atoms can be applied for a rather sensitive surface spectroscopy (MDS, see Boiziau 1981, Ertl 1986).

3.2.3 Doubly Charged Projectiles

The work of Hagstrum also includes extensive investigations on PE due to impact of doubly charged noble gas ions, e.g. on clean Mo (Hagstrum 1956e)

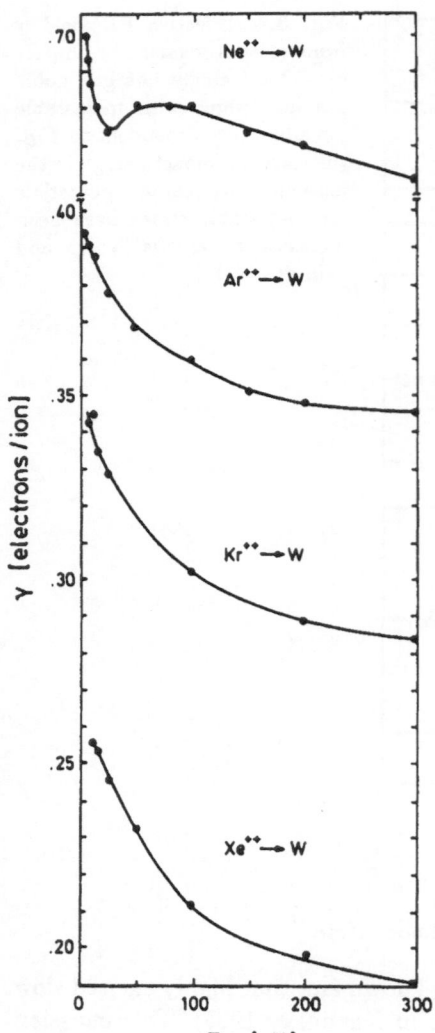

Fig. 3.5. PE yield γ for impact of doubly charged noble gas ions on clean polycrystalline tungsten, kinetic energy between 5 and 300 eV (Hofer 1983)

or W (Hagstrum 1954a). In the latter paper the qualitative idea has been presented that neutralisation of multicharged ions occurs in a series of roughly isoenergetic steps which with comparable probability would cause emission of electrons. Further support for this idea was provided from the corresponding ejected electron energy distributions (cf. Sect. 3.3), which for a given projectile charge state appeared to be included in those obtained for the next higher q value ("inclusion principle").

Figure 3.5 (Hofer 1983) shows the typical behaviour of PE yields for doubly charged noble gas ions impinging on a clean metal surface – the observed γ values are considerably higher than for singly charged projectiles and there is a pronounced decrease with increasing impact energy, in contrast to the

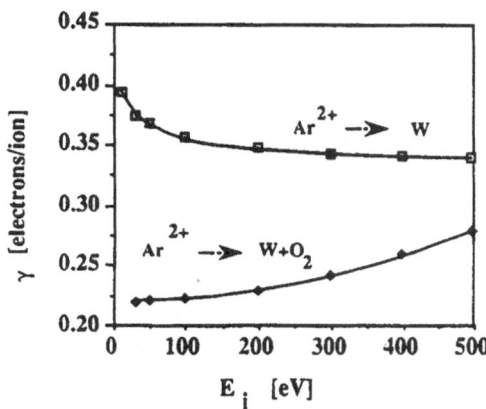

Fig. 3.6. Electron emission yield γ for Ar^{2+} impinging on clean and O_2-covered polycrystalline tungsten, respectively (Varga 1989)

situation for singly charged ions (see above). As already stated by Hagstrum (1954a), this is clear evidence for the importance of stepwise neutralisation-deexcitation mechanisms giving rise to the electron emission. The related ejected energy distributions (cf. Sect. 3.3) show particular structures from which a detailed information on the electronic transitions participating in the neutralisation processes of doubly charged ions near a metal surface can be derived.

Also for doubly charged projectiles the influence of electronegative surface adsorbates on the PE yield can be demonstrated as shown in Fig. 3.6 (Varga 1989), which compares results for a clean W surface (see also Fig. 3.5) and the same surface with O_2 coverage. In the latter case at low impact energy the yield is considerably smaller than for the clean surface, demonstrating the still important role of the target work function. Toward higher impact velocity, however, the yield increases due to an increasing KE contribution which is obviously more important than for the clean surface (cf. Hasselkamp 1991).

An influence of projectile excitation can be demonstrated also for doubly charged ions (cf. Fig. 3.7 from Varga and Winter 1978) and explained by the larger amount of available potential energy for the excited projectiles in comparison with the respective ground state ions.

3.2.4 Multicharged Projectiles

Based on the extensive experience with impact of singly and doubly charged projectiles (see above) and also multicharged ions (Hagstrum 1954a, $q \leq 5$ and Arifov et al. 1973, $q \leq 7$) on clean metal surfaces, a roughly linear relation between the projectile potential energy and the corresponding PE yields could be derived. This was also to be expected, if a step-by-step utilization of the available total potential energy takes place in an increasingly larger number of steps, each one involving either resonance neutralization/Auger deexcitation (Varga and Winter 1978) or multiple resonance neutralisation with subsequent autoionisation (Arifov et al. 1973), in summary giving rise

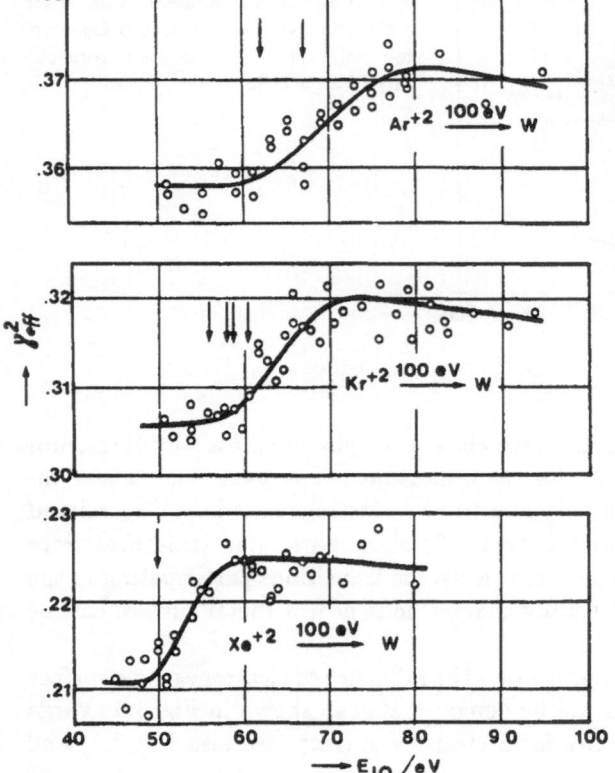

Fig. 3.7. Effective PE yield γ_{eff} from clean tungsten for impact of 100 eV doubly charged noble gas ion beams with metastable- ion admixture dependent on E_{IQ}, the electron impact energy in the ion source. Appearance potentials for metastable states have been indicated by arrows (Varga and Winter 1978)

to emission of a correspondingly large number of electrons. However, this linear dependence was found to break down as soon as projectiles in still higher initial charge states have been applied, e.g. N^{6+} or Ar^{9+} (Delaunay et al. 1985, 1987a, cf. Fig. 3.8a,b). In such cases the assumed partition of the total potential energy into relatively small parts for successive emission of electrons is apparently no more taking place.

Figures 3.9 and 3.10 show the impact energy dependence for bombardment with Ne^{q+} ($q \leq 7$) and Ar^{q+} ($q \leq 12$) on clean polycrystalline W (Varga 1987, Delaunay et al. 1987a). The now very pronounced decrease of the PE yields with increasing impact velocity can be qualitatively explained by a shortening of the available time interval between the onset of first possible resonance neutralisation transitions and the final impact on the surface (cf. Sect. 3.1), which limits the number of possible projectile deexcitation or autoionisation transitions. With still further increasing impact energy the yields tend to rise again as a consequence of the gradual onset of KE, cf. Chap. 4.

Several groups have performed similar studies for impact of multicharged ions on gas-covered metal target surfaces (e.g. Cano 1973, Arifov et al. 1976, Oda et al. 1988), the results of which support the above presented qualitative picture.

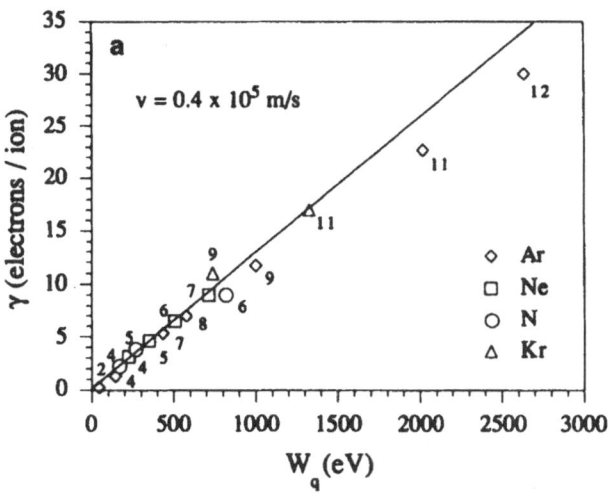

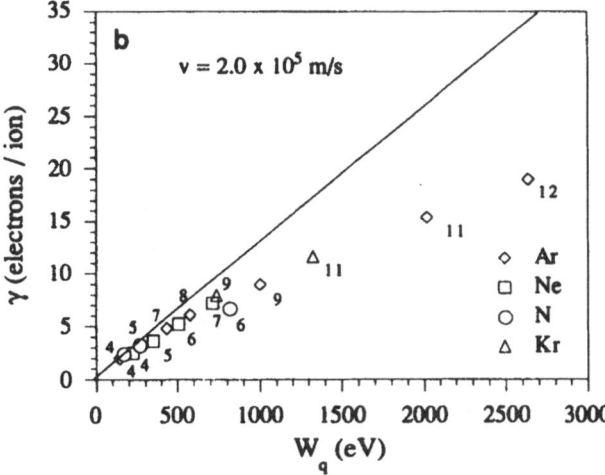

Fig. 3.8. (a) Total PE yields γ for impact of various ions on clean tungsten vs. total ion potential energy W_q at a given impact velocity of 0.4×10^5 ms^{-1} (Delaunay et al. 1987). The straight line is a best fit through the data points for low charge states, assuming a linear proportionality of γ vs. W_q. (b) Total PE yields γ for impact of various ions on clean tungsten vs. total ion potential energy W_q at a given impact velocity of 2.0×10^5 ms^{-1} (Delaunay et al. 1987). Straight line as in a). Note the progressive deviation of measured data with increasing q from the linear extrapolation according to a)

The clear deviation from the linear relationship of γ vs. total potential energy for high values of q (Fig. 3.8) can be understood in the following way. The resonance neutralisation processes as above referred to populate rather highly excited states in the approaching projectile, because of the small binding energies of the electrons in the solid as characterized by the surface work function. The subsequent autoionisation of such highly excited states involves in first instance a transition of one or more of the excited electrons into only

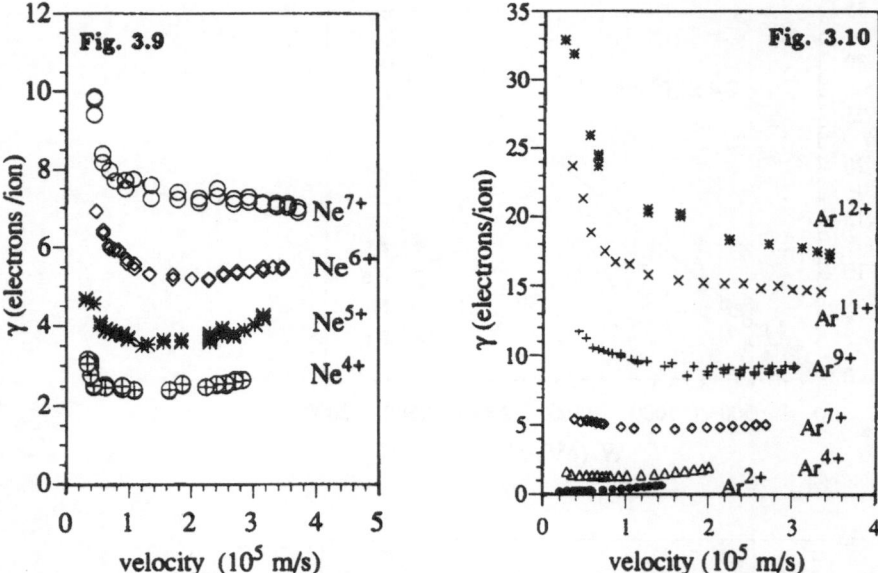

Fig. 3.9. Total PE yields γ for Ne^{q+} impact on clean polycrystalline tungsten (Fehringer et al. 1987)

Fig. 3.10. Total PE yields γ for Ar^{q+} impact on clean polycrystalline tungsten (from Fehringer 1987 and Delaunay et al. 1987a)

slightly deeper lying levels and therefore gives rise to emission of rather slow electrons only (Arifov et al. 1973, Andrä 1989). However, during the same time inner shell vacancies can develop in the projectile which in their later decay via intra-projectile Auger transitions will produce much faster electrons (see Sect. 3.3). In particular, these fast electrons are being produced if the original MCI carries already an open inner shell, because the corresponding vacancies involve much higher binding energies than the states primarily populated via first resonance transitions from the solid.

For highly charged ions the Auger transitions, which are recombining the inner shell vacancies and thus cause the emission of fast electrons, are starting to compete with radiative transitions as soon as the corresponding fluorescence yields have become comparably large.

Because of this transient formation of inner shell vacancies with subsequent decay via fast Auger electron emission, the related PE yields will fall short of expectations according to an extrapolated linear relationship from lower q-values for γ with the projectile total potential energy. The fast projectile Auger electrons (more details on this will be presented in Sect. 3.3) carry away a relatively large share of the totally available potential energy, which can thus not be completely used for the emission of a considerably larger number of slow electrons. Consequently, the onset of projectile inner shell vacancy formation has to be correlated with the levelling off in the PE yield – projectile potential energy dependence. Careful experiments with similar ion

– target combinations by de Zwart (1987) have provided further support for the above given explanations.

Finally it should be remarked, that the difference between "extrapolated" and actually observed PE yields for MCI carrying open inner shells becomes relatively more pronounced with higher impact energy (Fehringer et al. 1987, de Zwart et al. 1989). This can be understood from the comparably long lifetime of the transiently formed inner shell vacancies during the projectiles approach toward the surface. If the time of approach becomes shorter than the vacancy lifetime, the fast electron emission in front of the surface will become suppressed and the remaining potential energy will become dissipated inside the target bulk (see also Chap. 4).

The influence of target surface-coverage on PE yields has not yet been systematically investigated with multicharged projectiles. However, it is expected that deviations from the results obtained with clean metal surfaces will become relatively less important than for singly or doubly charged ions (cf. Sects. 3.2.1 and 3.2.3, respectively), because of the much higher density of projectile states into which electrons can be transferred, and the relatively smaller importance of a changing work function as compared to the now considerably larger total potential energies made available by the multicharged projectiles.

3.3 Experimental Results on PE – Electron Energy Distribution

3.3.1 Impact of Metastable Atoms

Electron emission by metastable excited atoms (in particular by He*$2s$ ^1S) with thermal kinetic energy has become a powerful method to study the properties of occupied valence electronic states at surfaces ("MDS"/Metastable Deexcitation Spectroscopy, see references in the review by Ertl 1986). It is much more sensitive to the surface conditions than e.g. photoelectron spectroscopy because the slow atoms cannot penetrate into the solid but start to interact already at a distance of several Å from the surface and therefore only with the outermost surface layer. The de-excitation of a metastable atom may occur by one of the following mechanisms.

1) AD, if the work function of the surface is low enough for RI becoming impossible. In this case the energy distribution of the emitted electrons reflects directly the S-DOS of the valence band. This situation occurs mainly with adsorbed alkali metals (Woratschek et al. 1985).

2) On clean and oxidized surfaces with relative large work functions, RI will occur as a precursor process for subsequent AN. RI will take place if the local work function exceeds the effective ionization potential of the metastable atom, in particular for the He*$2s$ levels (3.6 eV and 4.8 eV, respectively) in front of the surface. In such cases it has been shown by

Sesselmann et al. (1987a, b) that the energy distribution of the emitted electrons is a self-convolution of the S-DOS, similar as for the interaction of singly charged ions with the surface. Consequently, a deconvolution of the measured electron energy distribution is necessary to give information about the surface density of states. However, the low impact energy of the metastable atoms remains still an advantage in comparison to INS (Ion Neutralization Spectroscopy), see below.

3.3.2 Impact of Ions

Singly Charged Ions. Ion Neutralization Spectroscopy (INS), pioneered by Hagstrum was the consequent application of his theory of PE induced by ions with impact energy below 10 eV. Hagstrum published a large number of papers during the last three decades in which the electron emission induced by low energetic noble gas ions from clean and gas-covered metals as well as semiconducting surfaces has been extensively investigated (cf. reviews by Hagstrum 1956a, 1977, 1978).

Singly charged noble gas ions are neutralized mainly via AN processes and therefore the energy distribution of emitted electrons corresponds to a self-convolution of the density of states of the surface valence band as described in Sect. 3.3.1.

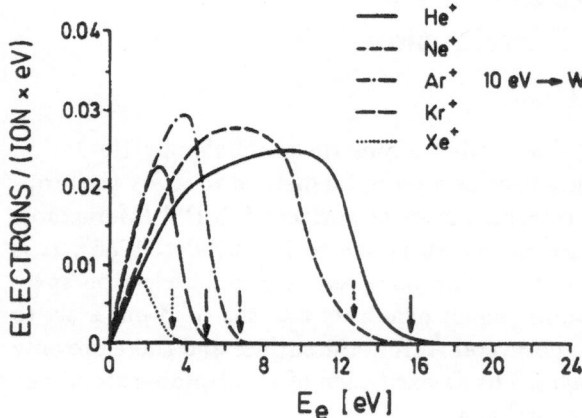

Fig. 3.11. Ejected electron-energy distributions for 10 eV He^+, Ne^+, Ar^+, Kr^+ and Xe^+ impact on clean polycrystalline tungsten. The arrows indicate the maximum possible energy assuming AN (Varga et al. 1982a)

In Fig. 3.11 from Varga et al. (1982a) measured electron energy distributions for impact of singly charged noble gas ions with 10 eV kinetic energy on clean tungsten are shown. The arrows indicate the maximum possible energy if AN is assumed. With increasing impact ion energy the electron energy distribution broadens (Figs. 3.12 and 3.13, from Hofer 1983). The fraction of low energy electrons decreases and additionally more electrons with large energy are produced.

When measuring the energy distribution (and also the yield, cf. Sect. 3.2) for PE, one has to take into account that a fraction of long-lived excited ions

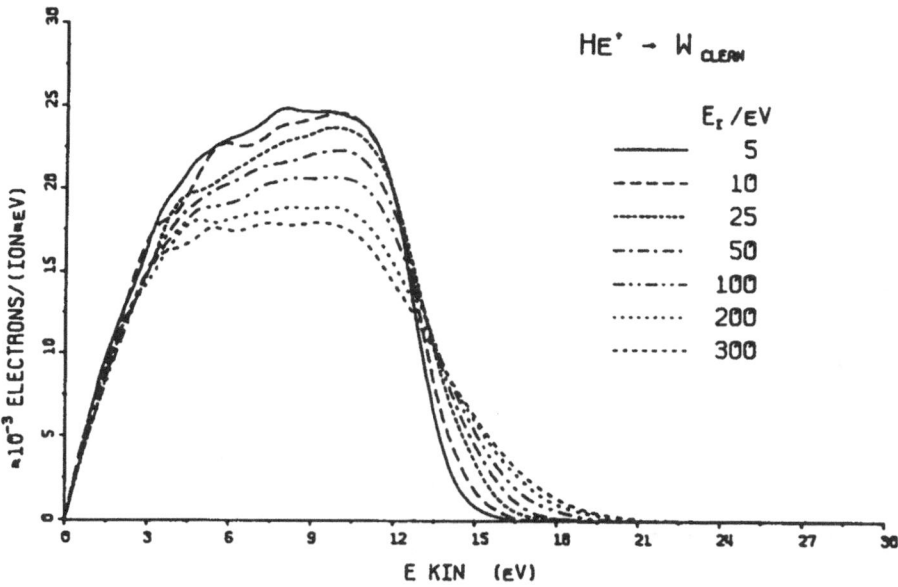

Fig. 3.12. Measured electron-energy distributions induced by neutralization of He⁺ on clean polycrystalline tungsten with impact energies between 5 and 300 eV (Hofer 1983)

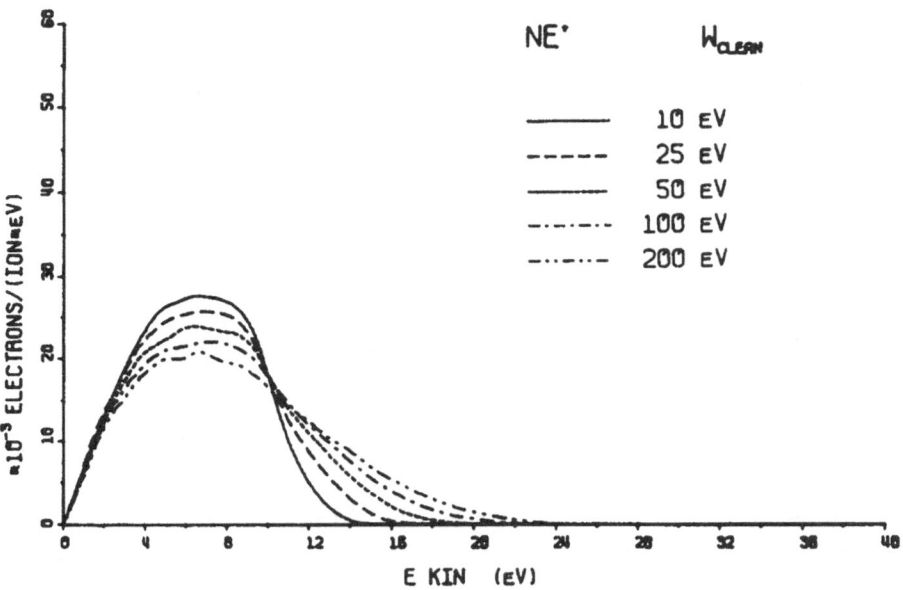

Fig. 3.13. Measured electron-energy distributions induced by neutralization of Ne⁺ on clean polycrystalline tungsten with impact energies between 10 and 200 eV (Hofer 1983)

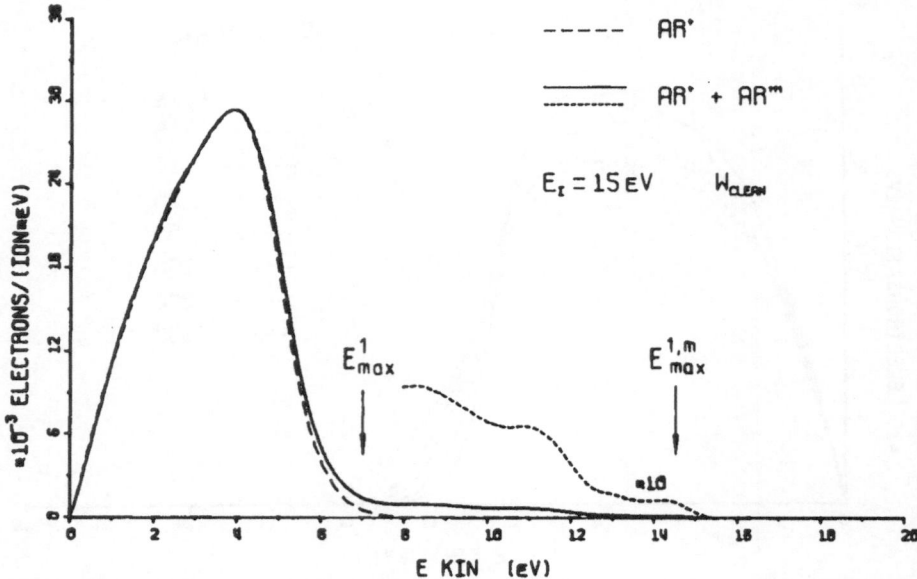

Fig. 3.14. Ejected-electron energy distributions for 15 eV Ar$^+$ impact on clean polycrystalline tungsten. Dashed line shows the result for ion beams without and the full and dotted lines with an admixture of metastabel excited ions, respectively (Hofer et al. 1983)

may be present in the ion beam. For the noble gas ions Ar, Kr and Xe several percent of metastable excited species can be given (Hagstrum 1956b, 1960, Varga and Winter 1978, Varga et al. 1981). If the potential energy of these metastable excited ions is large enough, a high energy tail in the electron energy distribution appears (see e.g. Fig. 3.14) and can be applied for the determination of the metastable ion beam fraction (Hofer et al. 1983b).

Doubly Charged Ions. It has been assumed that doubly charged ions are neutralized in a step-by-step manner, with each step involving similar processes as for impact of singly charged ions or excited atoms (Hagstrum 1954a,b, Arifov et al. 1973).

Electron transitions are most probable into highly excited projectile states which are energetically resonant with the surface valence band states (see Sect. 3.1). Such transitions may involve RN within the binding energy region of the valence band, but also AN from the Fermi edge to lower projectile states. It has been shown by quantum-mechanical calculations of Snowdon et al. (1986), that transition rates for RN and AN into states with the same binding energy are of comparable magnitude.

Woratschek et al. (1985) observed the conversion of metastable excited He*(^1S) atoms into the ^3S state in front of a surface as a precursor to further deexcitation into the He ground state. This is a strong indication that highly excited particles deexcite more probably via closely spaced states than directly into the ground state. They will undergo AD or AI directly into the atomic ground state only if no intermediate projectile states are available.

It has been demonstrated for normally incident 10 eV He^{2+} on Ni by Hagstrum and Becker (1973), that also double electron capture (e.g. a sequence of two RN processes) into autoionizing doubly excited states can take place for impact of doubly charged ion on a surface. Zeijlmans van Emmichoven et al. (1988) have observed similar AI processes at an impact energy of 1 keV with grazing incidence (He^{2+} with an incidence angle of 8° corresponding to a kinetic energy normal to the surface of about 20 eV).

Generally the situation is more simple for the neutralization of singly charged ions because the number of allowed transitions is rather limited. However, for doubly charged He, the still relatively small number of possibly involved excited levels permits to follow the neutralization steps in considerable detail. Guided by the similarity of observed electron energy distributions for impact of He^{2+} and $He^{+*}(2s)$ metastable ions, respectively (Hagstrum and Becker 1973) it can be argued that He^{2+} in a first step should become resonantly neutralized to the $He^{+*}(n = 2)$ state. Subsequently, a second RN into a doubly excited He^{o**} state takes place, with the latter than autoionizing to the $He^{+}(1s)$ ground state under ejection of a fast electron. The neutralization of $He^{+}(1s)$ takes place as known from INS (Ion Neutralization Spectroscopy) experiments (Hagstrum 1956a, 1977, 1978) via AN or RN/AD, depending on the work function of the target surface.

For heavier doubly charged noble gas ions a considerably larger number of excited states within the energy region of the surface valence band makes the situation rather more complex. However, measured electron energy distributions for impact of slow (< 100 eV) Ne^{2+}, Ar^{2+}, Kr^{2+} and Xe^{2+} still show significant structure (Varga et al. 1982a,b, Hofer 1983a).

As an example, in Fig. 3.15 electron energy distributions for impact of 10 eV Ne^{2+} on clean tungsten are shown together with relevant binding energy levels of Ne^{2+} and Ne^{+} in front of a tungsten surface. The numbered arrows indicate the different possible transitions, where the corresponding features in the electron energy distribution are marked with the same numbers. Especially the occurence of stepwise neutralization is demonstrated very nicely by the features around 6 eV, which are due to AN of the finally produced Ne^{+} ground state ions.

Since observed structures in the electron energy distribution are much more typical for the electronic structure of the projectile than the involved S-DOS, the target surface merely serves as a practically inexhaustible source of electrons.

With increasing impact velocity all distinct features of the electron energy distribution tend to disappear. For example, Figs. 3.16a,b,c show that the distribution of electrons emitted due to 200 eV Ar^{2+}, Kr^{2+} and Xe^{2+} impact has already become a smooth curve with one single maximum at low electron energy (Hofer 1983, Varga 1987, 1988). This may be caused mainly from level broadening, but also from nonadiabatic excitation of electrons in the solid

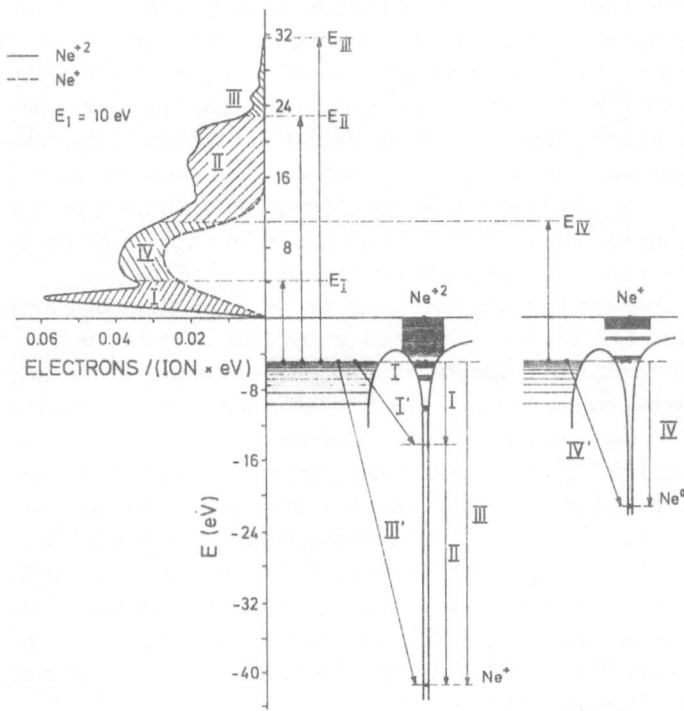

Fig. 3.15. Ejected electron energy distributions for Ne⁺ and Ne²⁺ impact on clean tungsten combined with electron energy diagrams for the W valence band, Ne⁺ and Ne²⁺. Arrows indicate some possible transitions and the corresponding maximum electron energy (Varga et al. 1982a)

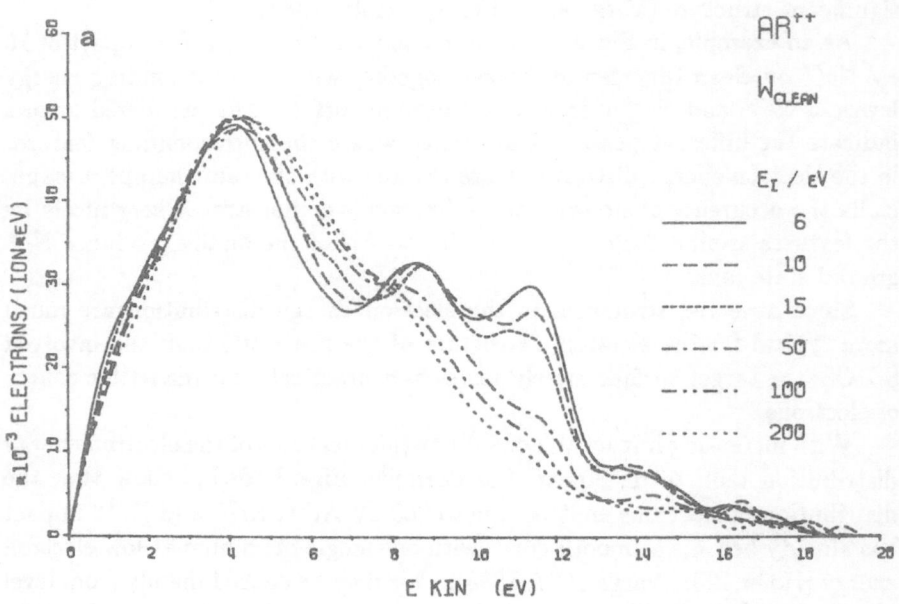

Fig. 3.16a. For caption see opposite page

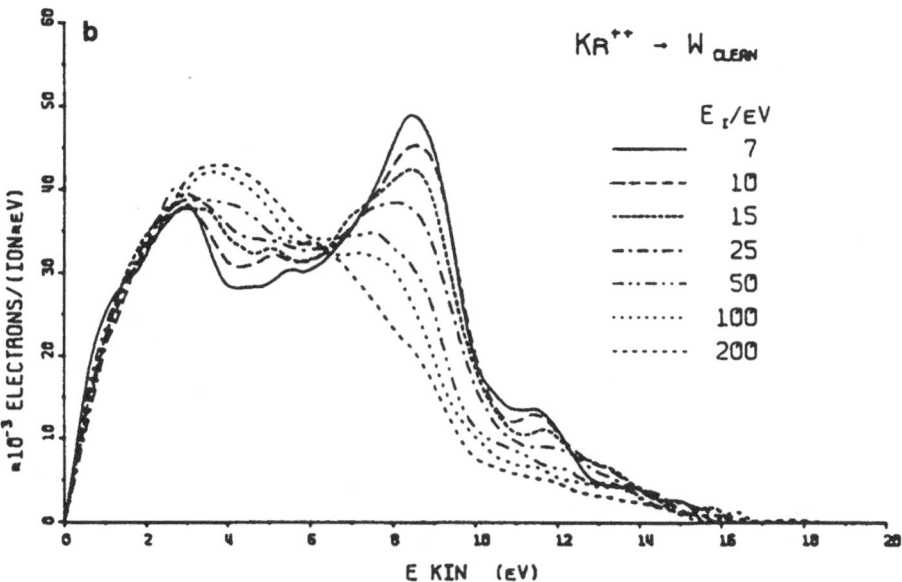

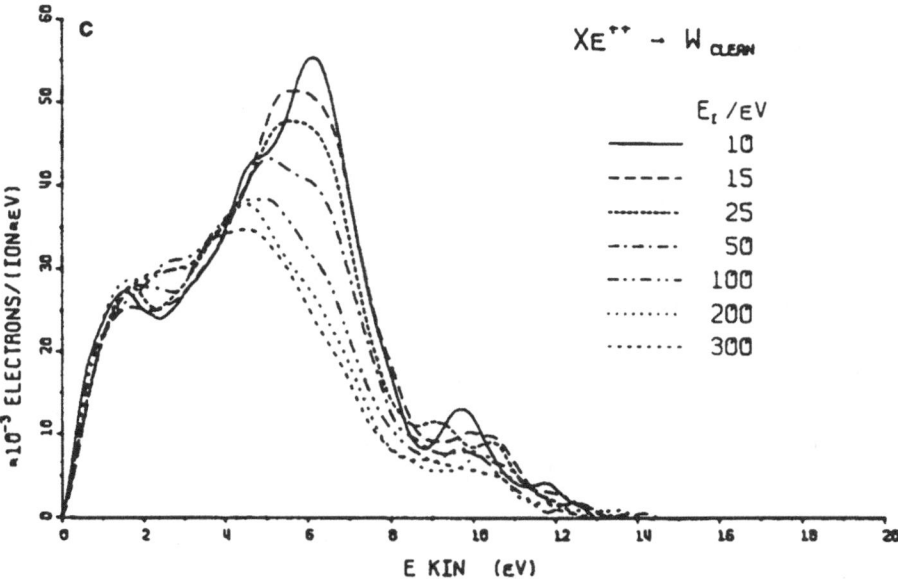

Fig. 3.16. (a) Ejected electron energy distributions for impact of Ar^{2+} on clean poly-crystalline tungsten at different impact energies (Varga et al. 1982a). (b) Ejected electron energy distribution for impact of Kr^{2+} on clean polycrystalline tungsten at different impact energies (Hofer 1983). (c) Ejected electron energy distributions for impact of Xe^{2+} on clean polycrystalline tungsten at different impact energies (Hofer 1983)

due to the ion motion as well as the fact that the neutralization is not yet finished if the projectile hits the surface (Hagstrum 1954a).

As the result of careful experiments with doubly charged keV He and Ar ions impinging at grazing angles on well defined single crystal surfaces, a quite complete explanation for the emitted electron spectra could recently be given by Zeijlmans van Emmichoven et al. (1988, 1990) and Wouters et al. (1989). The measured electron energy distributions have been reproduced by model calculations similar as described in Sect. 3.1, explaining the features above 3 eV almost quantitatively. Three types of processes are distinguished, i.e. AI, AD and AN. The model is based on the assumption that these different processes of charge exchange between the metal and the projectile can be distinguished from the processes of spontaneous electron emission. In addition it is assumed that the time evolution of the projectile-metal system during the interaction can be described by coupled rate equations. The electron energy distribution in this description is composed of contributions from the spontaneous decay of certain initial states of the system to certain final states, whereby these contributions involve finite energy widths because the electron emission can occur in a range of distances. Furthermore, it could be shown that the metal does not respond immediately to a change of the projectile charge state, but rather some delay in the replacement of surface electrons had to be taken into account.

Multiply Charged Ions. Hagstrum (1954a) as well as Arifov et al. (1973) have shown that the highest observable electron energy does not strongly increase with the available potential energy of the respective projectiles. For higher ion charge states only an increase of the low-energy part of the electron energy distribution is observed, and very little structure in this part of the energy distribution can be found. In Fig. 3.17 (Varga et al. 1982b) the energy distribution of electrons emitted due to impact of 30 eV $Ar^{q+}(q = 1, 2, 3)$ on clean tungsten clearly indicates the missing of high energetic electrons for Ar^{3+} (the arrows indicate the maximum possible electron energy if AN or RN/AD of Ar^{q+} into the ground state $Ar^{(q-1)+}$ are assumed). This is mainly caused by the increasing number (and therefore decreasing potential energy per step) of neutralization and deexcitation processes for the highly charged particle.

A drastic change has been observed for MCI carrying inner shell vacancies, because their deexcitation has to bridge at least one much larger step of potential energy. This has been demonstrated (de Zwart 1987, Zehner et al. 1986, Delaunay et al. 1987) for several ion species where high-energetic electrons in the electron energy distribution could be observed. Figure 3.18 (Varga 1988) shows electron energy distributions for impact of up to Ar^{9+} (an F-like ion), Fig. 3.19 (Delaunay et al. 1987b) for slow N^{6+} (H-like ion) impact on tungsten. A low-energy part of the distribution without any structure is seen together with a small percentage of much faster electrons originating from Ar-LMM and N-KLL Auger processes, respectively. The marks indi-

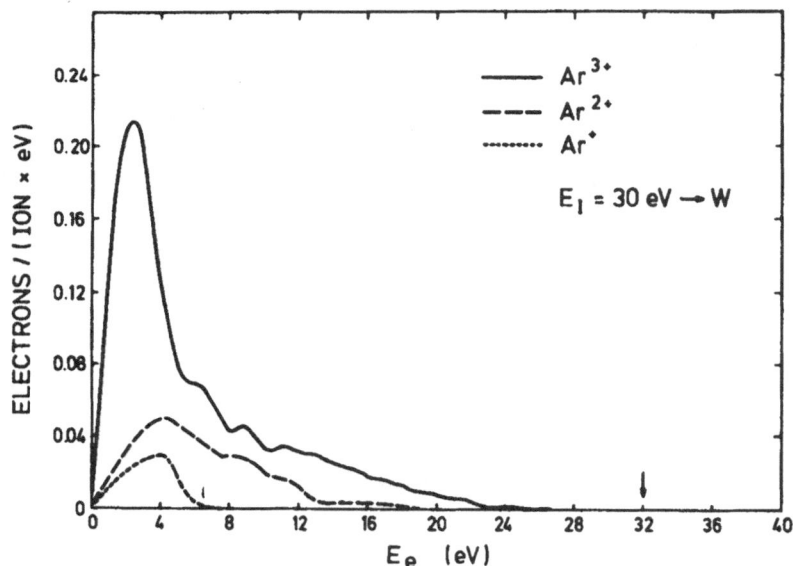

Fig. 3.17. Ejected electron energy distributions for impact of Ar^+, Ar^{2+} and Ar^{3+} on clean polycrystalline tungsten. The arrow indicates the maximum possible electron energy if assuming a single step for neutralizing Ar^{3+} into Ar^{2+} (Varga et al. 1982b)

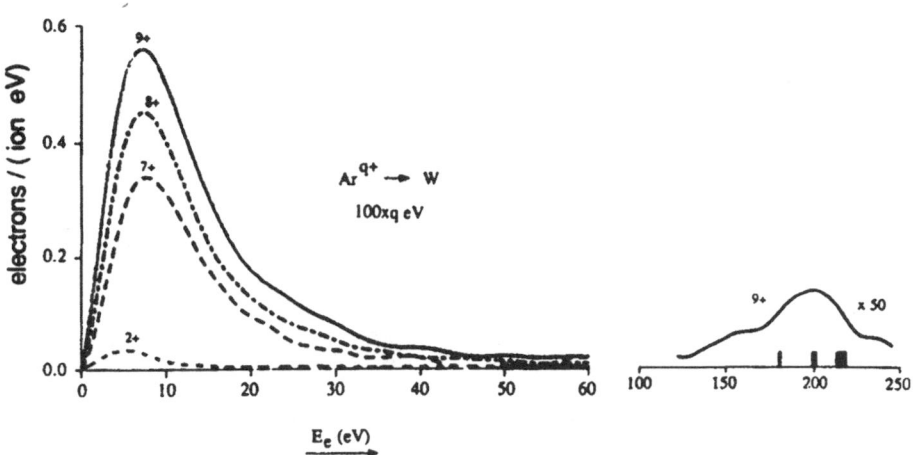

Fig. 3.18. Electron energy distributions for impact of Ar^{q+} on clean polycrystalline tungsten (impact energy of $100 \times q$ eV). Bars denote electron-induced LMM-Auger electron energies from Ar atoms (Varga 1988)

cate energies for emission of electron impact-induced Auger electrons. This evidence explains the neutralization of ions in charge states where inner shell electrons have been removed in the following way:

The outer shells of an ion approaching the surface are filled by RN and AN into highly excited Rydberg states, causing the formation of autoionizing

183

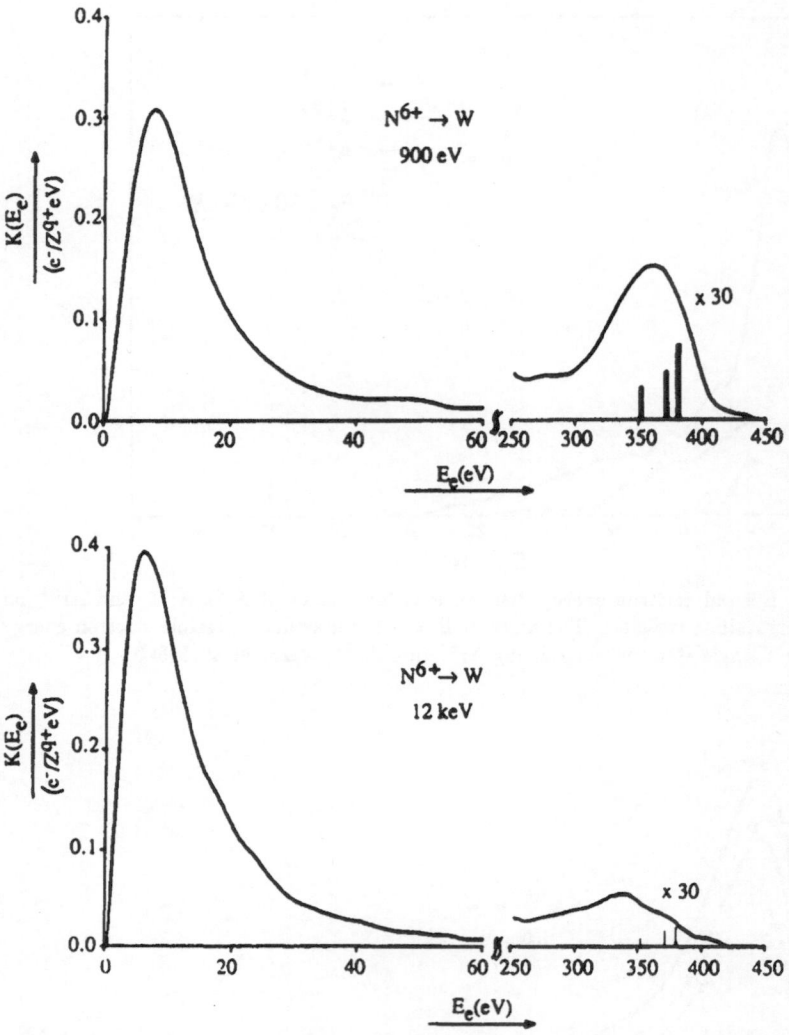

Fig. 3.19. Electron energy distributions for impact of N^{6+} on clean polycrystalline tungsten. Upper and lower curve for impact energies of 0.9 and 12 keV, respectively. Bars denote electron induced KLL-Auger electron energies from N atoms (Delaunay et al. 1987b)

states nearly resonant with the Fermi edge. Deexcitation occurs via AD and AI processes bridging as small energy gaps as possible (Andrä 1989), whereas an inner-shell vacancy in the projectile will survive these neutralization steps since its lifetime can be considerably larger than the time needed for the above mentioned AI processes. Finally, the inner shell vacancy becomes filled by a single intra-atomic Auger process emitting the observable fast electrons.

Further considerations show that for ions with low impact energy (≤ 1 keV) almost each Ar^{9+} and each N^{6+} ion should produce one LMM – or KLL – Auger electron, respectively, before surface impact (Delaunay et al. 1987b).

Carefully performed energy analysis of LMM-Auger electrons emitted due to Ar^{9+} ion impact at different energies and angles demonstrated their Doppler shift, which proves that the fast electrons are emitted by a projectile still on its way toward the surface (de Zwart 1987). With increasing impact velocity the yield of the high energy electrons (N-KLL or Ar-LMM electrons etc.) decreases drastically because the respective Auger emission cannot take place before the ion hits the surface (de Zwart 1987, de Zwart et al. 1989, Delaunay et al. 1987a,b).

Köhrbrück et al. (1991) has also investigated the Doppler shift of Auger electrons emitted due to scattering of 90 keV Ne^{9+} on a gas-covered copper surface. In this case the inner shell vacancies giving rise to the Auger electron emission could be shown to survive the scattering event. The same information has been obtained from measurements of the projectile charge states after their scattering on clean tungsten (de Zwart et al. 1985). For 20 keV Ne^{9+}, Ar^{9+} and Kr^{9+} primary ions a fraction of triply charged ions turned up after scattering, which for projectiles prepared in lower charge states could not be observed. From this result it can be concluded that the transiently formed inner shell vacancies survive the scattering event and only thereafter give rise to the Auger electron emission, thus finally increasing the charge state of the scattered partially or fully neutralized projectile by one.

It should also be mentioned that for N^{5+} (He like) a high energy peak in the electron energy distribution was observed, similar to that for N^{6+} (see Fig. 3.20). This can be explained due to an admixture of metastable excited ions and made it possible to evaluate the fraction of long lived excited particles resulting from the applied ion source (Delaunay et al. 1987b).

If an ion comes closer to the surface while still containing an inner shell vacancy, electrons from the target atom core can apparently fill such a vacancy. The resulting core holes in the target atom deexcite by emitting characteristic target Auger electrons, as has been observed for impact of fast MCI (de Zwart 1987). It is obvious that inner shell vacancy transfer from the projectile to the target atom can take place only in energetic collisions because of the necessarily overlapping core states. Therefore, target Auger lines have been observed only during the interaction of fast ions (to assure sufficiently small impact parameters) under grazing incidence (to avoid deep penetration into the target).

Recently, Folkerts and Morgenstern (1990) determined electron energy distributions for impact of slow hydrogen-like MCI (C^{5+}, N^{6+}, O^{7+} and Ne^{9+}) on clean tungsten. These authors could identify Auger electron emission not only from $KL \geq L-$, but also $LM \geq M$ transitions. By comparing the relative importance of both contributions they concluded that in the course of projectile neutralisation the projectile L shell is not exclusively populated via cascading from higher levels, but in addition some direct (QRN?) transitions into the very tightly bound L shell have to be assumed. Further investigations along these lines might yield more quantitative results for the sequence and duration of neutralisation processes giving rise to the observed PE.

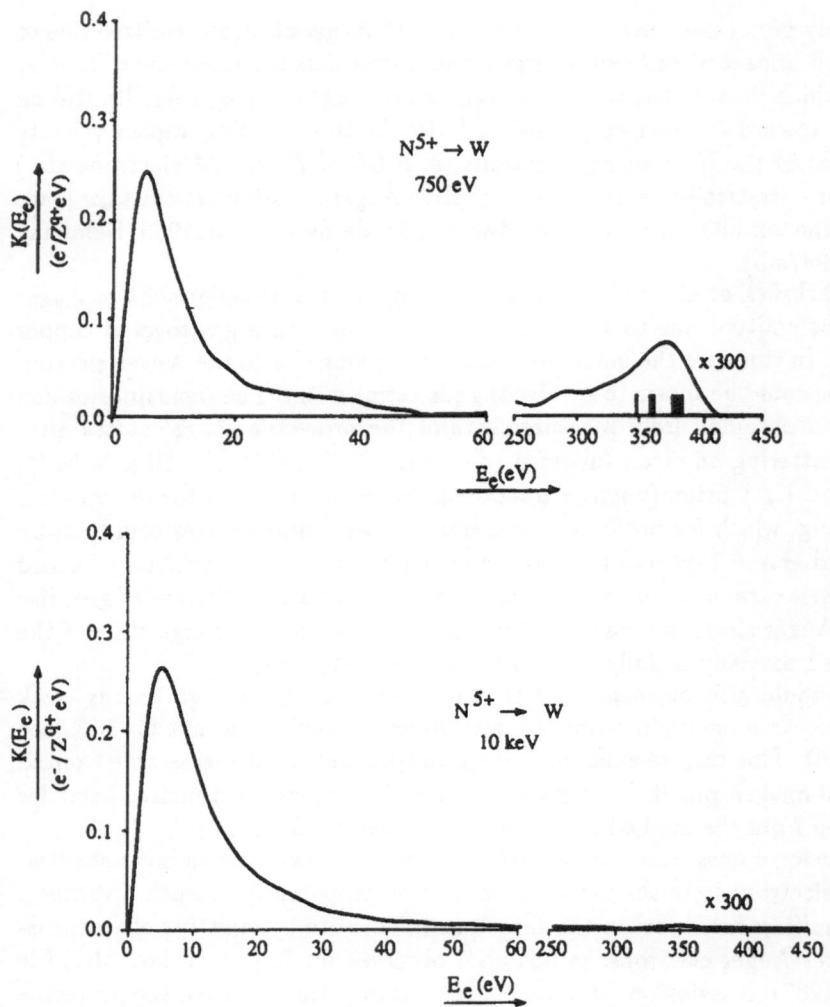

Fig. 3.20. As for Fig. 3.19, for N^{5+} ions with small metastable admixture at impact energies of 0.75 and 10 keV, respectively (Delaunay et al. 1987)

Influence of adsorbates. If the work function of the surface is changed, also the potential energy which is converted into the emission of PE electrons becomes changed. Similar as in the case of singly charged projectiles (cf. Fig. 3.3) this can be seen in Fig. 3.6 (Varga 1989) for the total electron yields measured for Ar^{2+} impinging on clean and oxygen-covered tungsten. Oxygen is known to increase the work function, to decrease the density of states at the Fermi energy and produce resonant states about 5 eV below the Fermi energy. For all these reasons, the electron yield is decreased if oxygen is adsorbed on a metal surface especially at low impact energy where PE dominates. At higher energy where KE becomes more and more important a strong increase in the total yield can be observed.

The change in the target electron density of states (the S-DOS of the metal is screened very effectively by the adsorbate) causes a drastic change of the emitted electron energy distribution (Hofer and Varga 1984, Sesselmann et al. 1987b).

Last but not least, at low coverage an adsorbate-induced S-DOS can change the probability of some precursor resonance transitions such that completely different AD processes will take place after adsorption as compared for the clean metal surface. Alkali adsorbates which decrease the work function drastically, will increase the probability of RN with following AD instead of pure AN, which is observed for clean metal surfaces.

For all doubly charged noble gas ions a change to an electron energy distribution without any structure, similar to those being typical for kinetic emission, has been observed for heavily contaminated surfaces. In Figs. 3.21a,b,c the influence of adsorption of oxygen on the electron energy distribution induced by the impact of 15 eV Ar^{2+}, Kr^{2+} and Xe^{2+} respectively is shown. As a qualitative statement, increasing adsorption changes the electron energy distribution in a similar way as increasing the impact velocity, with the underlying reasons being however completely different.

Because of similar reasons as already discussed for the PE yields (cf. Sect. 3.2.4), also the electron energy distributions should become less subject to adsorbates on the target surface if higher projectile charge states are regarded. However, systematic investigations of this aspect have not yet been carried out.

3.4 Experimental Results on PE – Electron Emission Statistics

Figure 3.22 shows schematically a recently developed setup for measuring electron emission statistics ("ES") down to considerably smaller ion impact energies than formerly accessible ($E \geq 50q$ eV, cf. G. Lakits et al. 1990a). The projectile ions can be decelerated just in front of the target, which is situated inside a highly transparent conical electrode serving for deflection of emitted electrons toward a solid state detector behind the target. The electrons are accelerated into this detector with an energy of up to 25 keV by means of a three cylinder lens, with all other features the same as for an earlier developed detector system based on similar working principles (Lakits et al. 1989a, Aumayr et al. 1991). With the here described setup one can also determine very small electron yields in the following way. Independent from the ion deceleration next to the detector, the ion beam intensity can be reproducibly attenuated by a variable factor of up to 10^6 with a magnetic quadrupole lens in the ion beam line far upstream of the ES detector. With ion currents of the order of nA and for $\gamma \geq 0.1$, the latter can be obtained by current measurements in the usual way (cf. Lakits et al. 1989a). If the ion current is strongly attenuated to a flux of typically 10^3 particles/s or less,

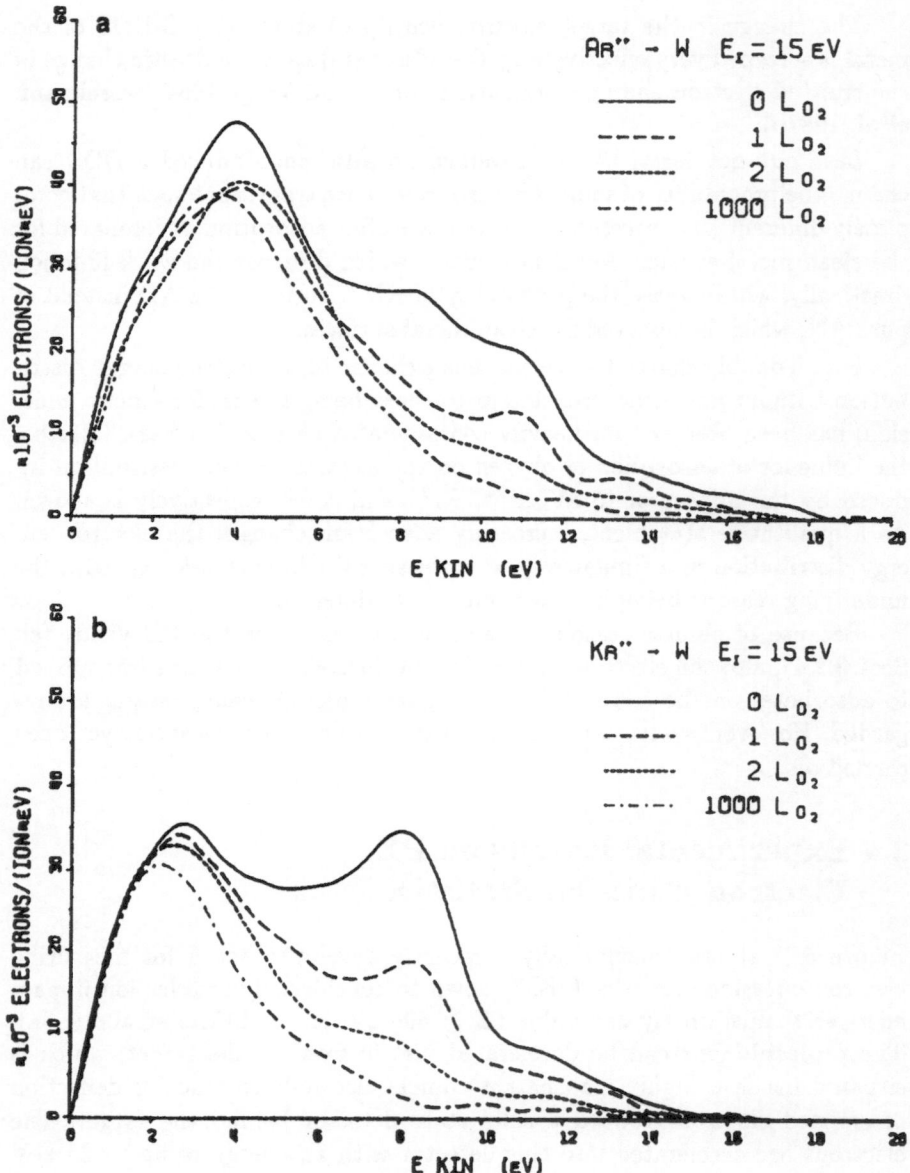

Fig. 3.21. For caption see opposite page

ıt can be precisely measured by counting all the electrons emitted from the target and taking into account the already determined value of γ. If than by an appropriately designed ion beam transport system the ion deceleration can be performed without any ion losses, a once set ion flux can be kept constant if decelerated. By measuring electron fluxes for various ion impact energies involving the known ion fluxes, the values for γ corresponding to other impact energies can be precisely determined even at such low values of E_i where a

188

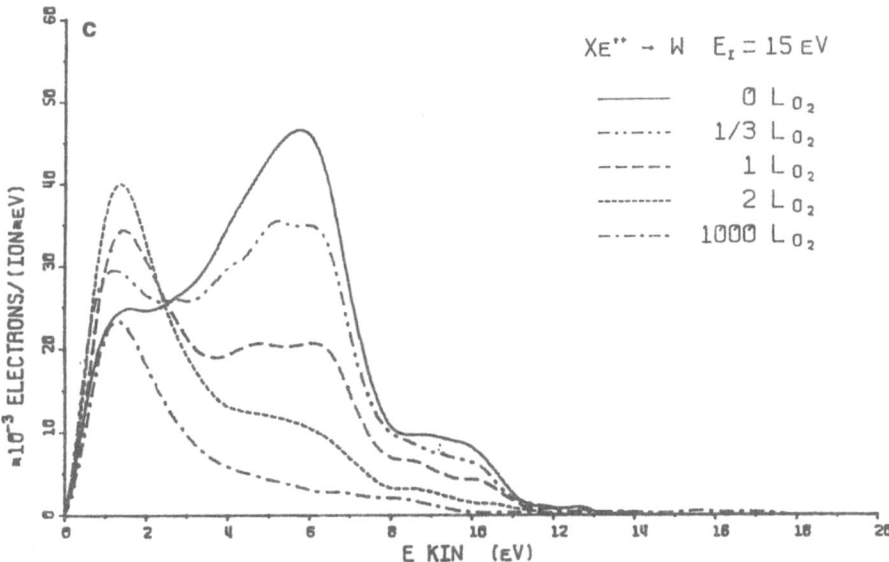

Fig. 3.21. (a) Measured ejected electron energy distributions for 15 eV Ar^{2+} impact on clean and oxygen-covered polycrystalline tungsten. The O_2-dose is given in Langmuir (1L = 10^{-6} torr s) (Hofer and Varga 1984). **(b)** As for Fig. 3.21a, but for Kr^{2+} (Hofer 1987). **(c)** As for Fig. 3.21a, but for Xe^{2+} (Hofer 1987)

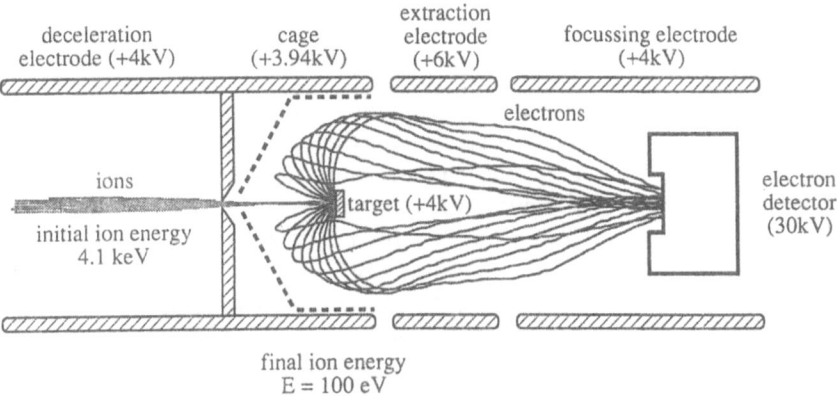

Fig. 3.22. Experimental setup for measuring electron emission yields and statistics for ion impact energies $E_i \leq 30q$ eV (Lakits et al. 1990a)

current-related γ measurement is no more feasible because of a far too small electron yield.

Figure 3.23 shows a series of raw ES "spectra" for impact of Ne^+ ions on clean gold (we remark that Ne^+ has no long-lived excited state and therefore only ground state ions can be involved), with the impact energy varied from 100 eV up to 16 keV. As clearly visible, at the lowest impact energy only one electron can be ejected, whereas with increasing E_i also the probabili-

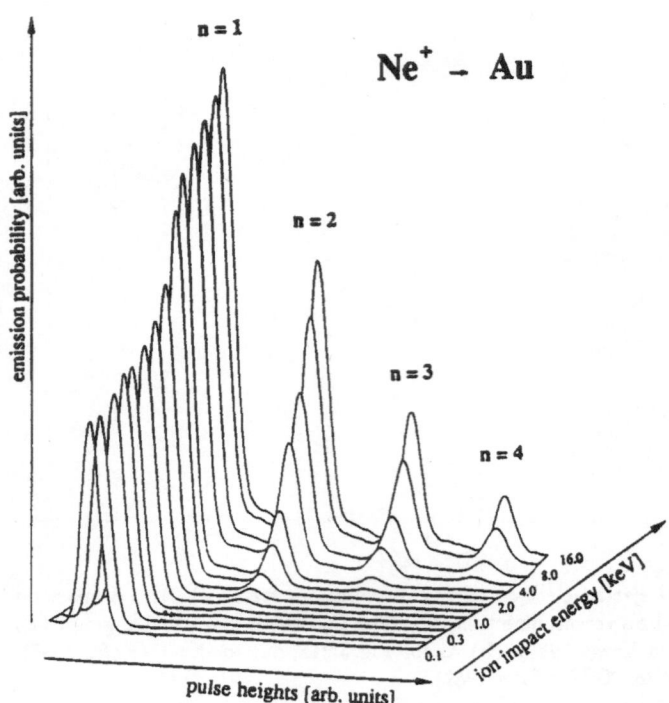

Ne$^+$ → Au

n = 1

n = 2

n = 3

n = 4

emission probability [arb. units]

16.0
8.0
4.0
2.0
1.0
0.3
0.1

ion impact energy [keV]

pulse heights [arb. units]

Fig. 3.23. Raw experimental data for electron emission statistics from Ne$^+$ impact on clean polycrystalline gold ($0.1 \leq E_i \leq 16$ keV; Winter et al. 1991)

ties for emission of 2, 3, etc. electrons gradually are coming up. Apparently, at low impact energy the PE mechanism contributes exclusively to the total electron yield by ejecting only one electron, although the available potential energy ($W \leq 21{,}6$ eV) would suffice for emission of two or even three electrons. This remarkable behaviour was further investigated by comparing apparent electron yields and corresponding ES for impact of equally fast Ne$^+$ and Na$^+$ ions, see Figs. 3.24a,b. For Na$^+$, which cannot provide a sufficiently high potential energy to cause PE from clean gold, the total yield is exclusively due to KE and gradually disappears toward low E_i, whereas for Ne$^+$ it approaches a fairly constant value corresponding to the there also involved PE contribution. However, for both ion species the ratio W_2/W_1 disappears at a similar low value of E_i (cf. Fig. 3.24b), which shows that in both cases a second electron is produced only after surpassing a similar impact energy, which thus clearly corresponds to the KE threshold being similar for both ions (see also Chap. 4).

Our here discussed results already demonstrate a fundamental difference of ES for PE and KE processes, respectively. PE involves well defined transitions between the electronic states in target and projectile and therefore also gives rise to emission of a well defined number of electrons. KE, on the other hand, results from statistical dissipation of projectile kinetic energy among a

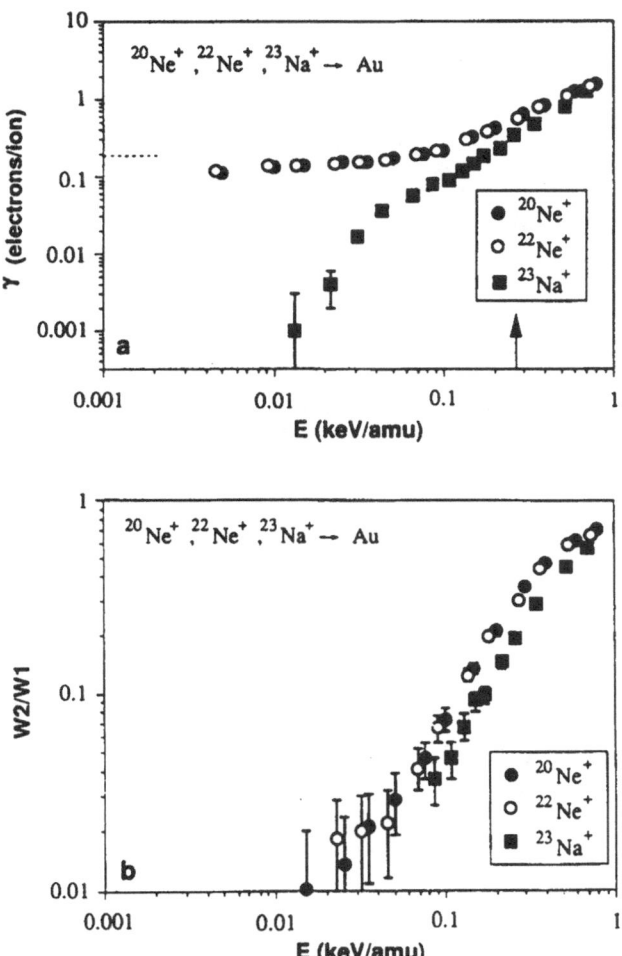

Fig. 3.24. (a) Total electron yields derived from ES measurements vs. ion impact energy for bombardment of clean gold by singly charged Ne- and Na ions; (b) Ratios of probabilities for emission of respectively two and one electrons in collisions as defined under a), vs. ion impact energy (Winter et al. 1991)

larger number of target electrons which are facing comparable chances of being ejected. The number of these electrons increases with transferred kinetic energy and thus E_i.

Similar data as for impact of Ne⁺ have been obtained with a variety of other projectiles, e.g. for H⁺ and He⁺, cf. Figs. 3.25a,b and for Ar⁺ and Xe⁺, cf. Figs. 3.26a,b (Lakits et al. 1990a). The very interesting result that for singly charged ground state ions the PE process can deliver only one electron irrespective of the potential energy carried by the ions toward the surface, points to a small interaction of target electrons (or, in other words, weak "electron correlation") at the Fermi edge. For clean gold this correlation is known to be rather weak and thus the potential energy offered by the incoming

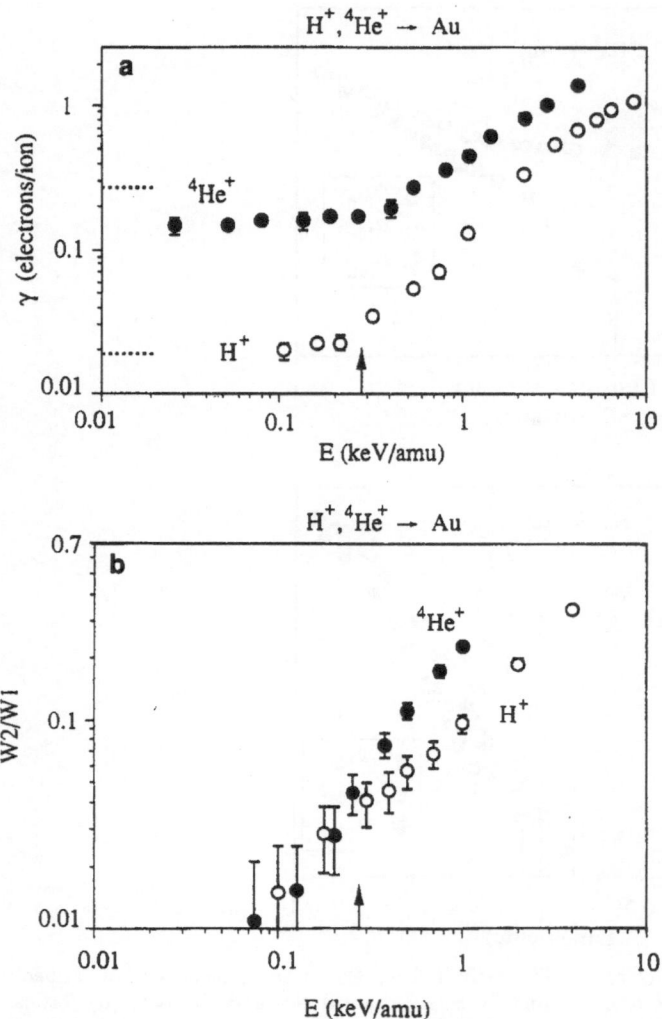

Fig. 3.25a,b. As for Figs. 3.24a,b, but for impact of H⁺ and He⁺ (Lakits et al. 1990a)

ion can be transferred by means of Auger type PE processes practically only to one electron in the solid.

Similar studies have been made with projectiles of given species and kinetic energy but in different charge states. Figs. 3.27a,b show ES for Ar ions in charges states $1 \leq q \leq 4$, which have been obtained after folding out from the actually measured ES those obtained for the equally fast neutral projectiles (for more details see Lakits and Winter 1990). Obviously, in the particular case of Ar^{2+} this deconvolution works only well at projectile impact energies below 16 keV, which we interpret in the following way (see also Sects. 3.1–3.3):

At low ion impact energy there is sufficient time for complete neutralization of the projectile until its impact on the surface, thus giving rise to

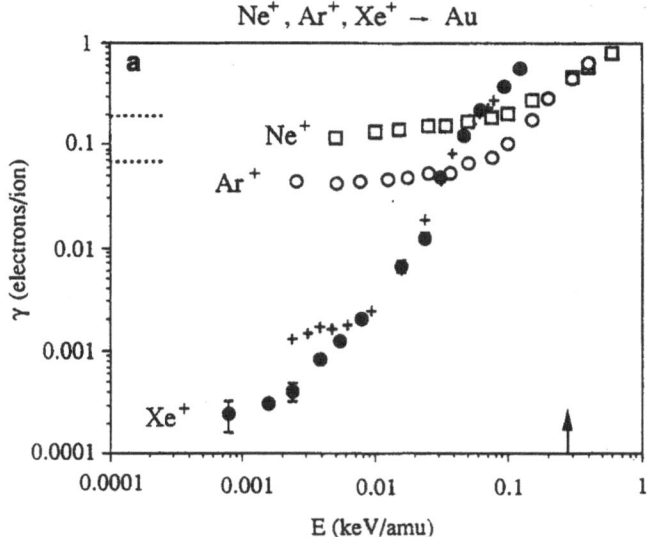

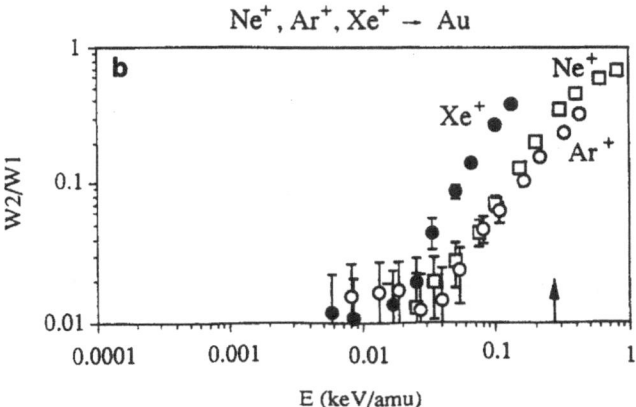

Fig. 3.26a,b. As for Figs. 3.24a,b, but for impact of Ne[+], Ar[+] and Xe[+] (Lakits et al. 1990a)

an equal KE contribution as for originally neutral projectiles. In such cases the PE and KE contributions should remain mutually independent and their respective ES will therefore be separable. However, at higher impact energy (in the case of Ar^{q+} for $q \geq 3$ at about 16 keV, corresponding to an impact velocity of ca. 3×10^5 m/s, cf. Lakits and Winter 1990) the projectile will not yet become completely neutralized until its impact on the surface, and the correspondingly initiated KE will therefore differ from that for originally neutral projectiles with the same impact energy. Under these circumstances the deconvolution can no more yield meaningful results, since the processes of PE and KE are not only interrelated, but the KE will also depend via impact energy on the original projectile charge state q (see further discussion of this point in Chap. 4).

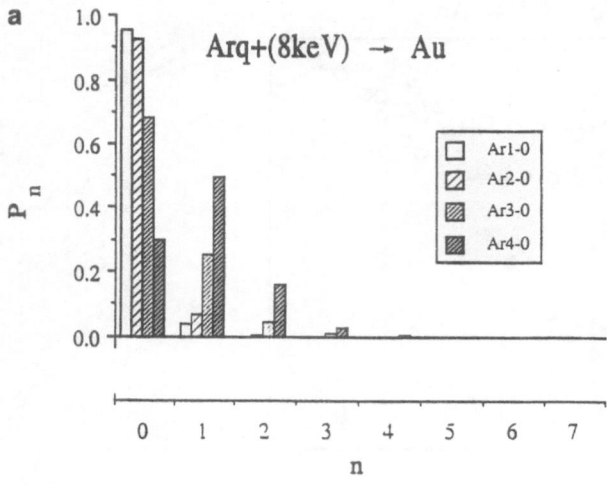

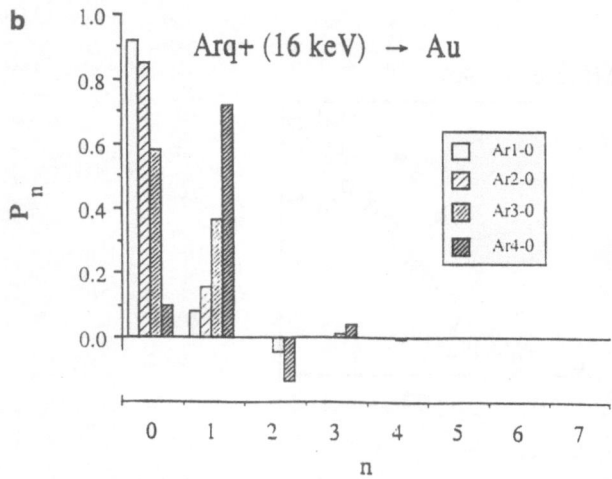

Fig. 3.27. (a) Electron emission statistics (ES) for impact of 8 keV Ar^{q+} ($q = 1 \div 4$) on clean gold, after deconvolution from ES for the equally fast neutral Ar projectiles; (b) As (a), but for 16 keV impact energy (Lakits 1990)

Figures 3.28a,b show similar results for impact of Ne^{q+} ($q = 0,1,2,3$) on clean gold (Lakits 1990, Winter 1991). In this case the breakdown of the above described ES deconvolution is observed at a Ne^{q+} impact energy above 6 keV, corresponding to a projectile velocity of ca. 2×10^5 m/s, because of the same reasons as already explained for the case of Ar^{q+} projectiles.

A final remark should be devoted to the observation that even for the lowest accessible MCI impact energies it has not been feasible to unfold ES

Fig. 3.28. (a) ES for impact of Ne^{q+} ($q = 0 \div 3$) on clean gold vs. impact energy; (b) As for Fig. 3.27a, but for Ne^{q+} ($q = 1 \div 3$; Winter 1991)

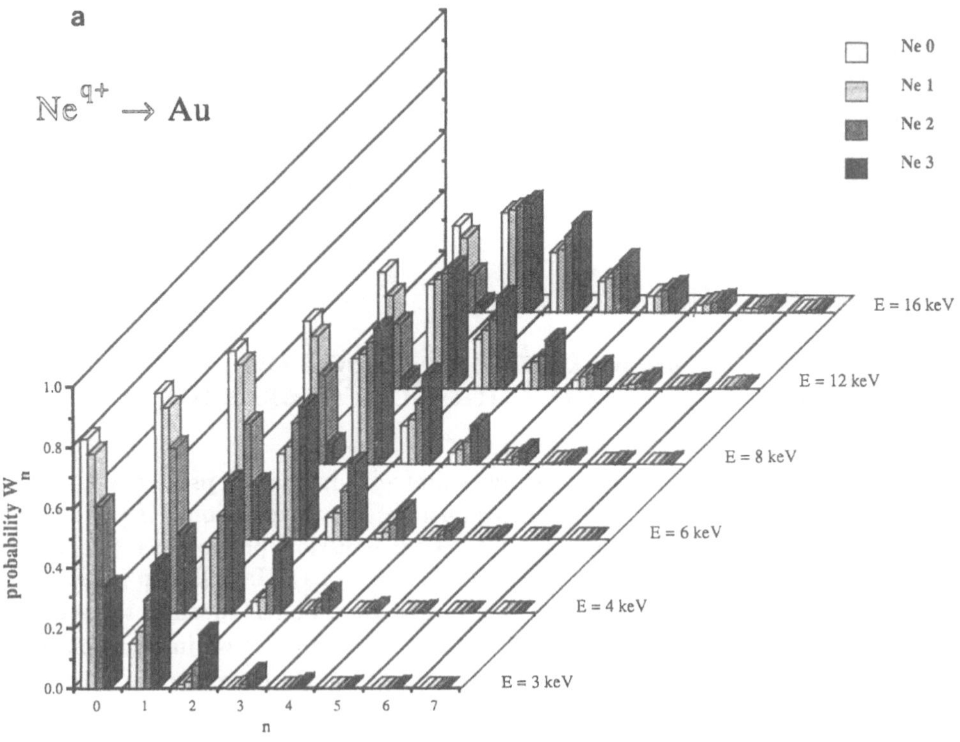

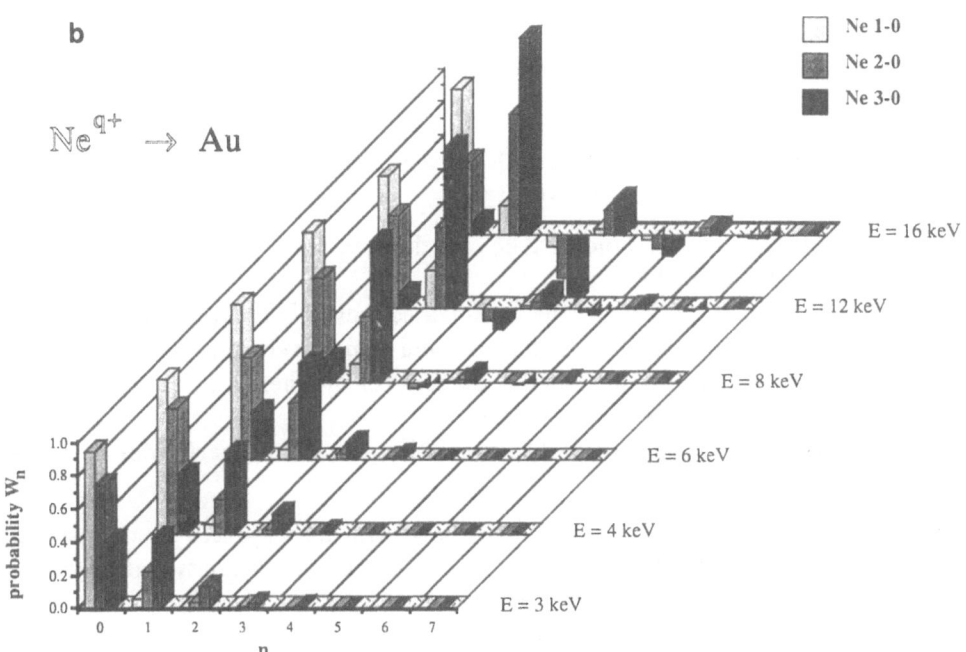

a
$Ne^{q+} \rightarrow Au$
Ne 0
Ne 1
Ne 2
Ne 3
E = 16 keV
E = 12 keV
E = 8 keV
E = 6 keV
E = 4 keV
E = 3 keV
probability W_n
1.0
0.8
0.6
0.4
0.2
0.0
0
1
2
3
4
5
6
7
n

b
$Ne^{q+} \rightarrow Au$
Ne 1-0
Ne 2-0
Ne 3-0
E = 16 keV
E = 12 keV
E = 8 keV
E = 6 keV
E = 4 keV
E = 3 keV
probability W_n
1.0
0.8
0.6
0.4
0.2
0.0
0
1
2
3
4
5
6
7
n

Fig. 3.28. For caption see opposite page

obtained for a given projectile charge state q from the ES for the same projectiles prepared in lower original charge states $q' = q - 1$, if $q' \geq 1$.

This is clear evidence that the multiple resonance neutralization transitions which initiate the PE processes (cf. Sects. 3.1–3.3) are producing fully neutralized projectiles at an early stage of the whole PE process, which then via autoionization decay under electron emission instead of a step-by-step neutralization along the sequence of charge states decreasing from the initial q toward $q = 0$.

4. Slow Particle-Induced Electron Emission at the Kinetic Threshold

We have mentioned at various instances that with increasing projectile impact energy in the presence of PE the mechanism of kinetic emission (KE) will also contribute to the than "apparent" electron emission phenomena (electron yield, energy distribution, statistics, cf. Sects. 3.2–3.4). Hasselkamp (1991), in the present volume, provided an exhaustive account on KE to which the reader may resort for further relevant informations. At this place we just summarize that in KE processes one has to distinguish between the three steps of electron production *inside* the target bulk, the diffusion of electrons toward the surface and finally their escape across the surface barrier into vacuum. Whereas the second and third steps can in first approximation be regarded as decoupled from step one and should thus not strongly be correlated to any property of the primary projectiles, step one will not only depend (as usually assumed) on the projectile species and impact velocity, but also its detailed structure, charge state, etc.

In this chapter we will deal with three questions connected with the combined action of PE and KE processes.

1) Where is the threshold of KE, especially in the presence of a non-negligible PE contribution?
2) How far can PE and KE be assumed as mutually independent?
   For these two questions, to be answered respectively in Sects. 4.1 and 4.2, we will revisit Sect. 3.4 and discuss some of the therein already presented results in somewhat more detail.
3) How does KE depend on the projectile properties apart from its mass and impact velocity?

In Sect. 4.3 the first step of KE (see above) will be inspected by comparing KE induced by projectiles of given species and impact velocity but in different charge states and chemical binding situations (viz. atomic vs. molecular projectiles, socalled "molecular effects", see also Hasselkamp 1991, Chap. 5).

4.1 The Precise Threshold of Kinetic Emission

If there is no significant PE contribution to be expected, as in the case of neutral projectiles or ions with too small potential energy (like the alkali ions, cf. Fig. 3.24a), the onset of KE will be clearly apparent from the course of γ vs. impact energy/velocity (cf. Waters 1958, Petrov 1960, Arifov et al. 1962, Medved et al. 1963). With PE also working, however, the course of the total electron emission yield does in general no more permit a precise judgment on the relative importance of PE and KE contributions, and in particular on the location of the KE threshold. As an example, Fig. 3.26a shows γ data for impact of Ne^{+}-, Ar^{+}- and Xe^{+} ground state ions on clean gold down into the low impact energy region of exclusive PE. The there expectable PE contributions have been estimated from semiempirical calculations (Kishinevskii 1973) and indicated by horizontal dashed lines. The potential energy carried by Xe^{+} ($W \leq 12,1$ eV) is apparently too small for a measurable PE contribution. Figure 3.26a also contains earlier measured data for Xe^{+} from Alonso et al. (1986), which toward low E deviate from the present results for reasons discussed by Lakits et al. (1990a).

We remark that these measurements extend to very low values of γ which so far could not be determined with a comparable accuracy. However, as far as the KE threshold is concerned, Fig. 3.26a does not offer much help, but Fig. 3.26b shows ratios of the relative ES probabilities for emission of $n = 2$ and 1 electrons, respectively. From these data the KE threshold impact velocity can be clearly identified since (at least for a clean gold surface) PE involves the emission of one electron only (cf. Sect. 3.4). A common concept for the KE threshold assumes that the minimum kinetic energy transfer in head-on collisions between the projectile particle and some quasi-free metal electrons has to surpass the metal surface work function (Alonso et al. 1980). The corresponding threshold impact velocity has been marked in Figs. 3.26a,b by a vertical arrow. As clearly visible, non-negligible KE contributions can be observed well below this "conventional" KE threshold, with contributions becoming somewhat more important for the heavier projectiles. We relate this "low E" KE fraction to close collisions of projectiles with individual target particles at the uppermost surface layer, causing electron emission from transiently formed autoionizing quasimolecules. Calculation of the probability for such processes by considering the impact energy-dependent distance of closest approach between the collision partners may lead to a more quantitative understanding of such processes, which should become even more important for contaminated metal surfaces, where however an interrelation with ion-induced desorption processes might take place.

Summing up, in the presence of PE from clean metal surfaces due to impact of singly charged ions we can place the threshold impact velocity for onset of KE where the corresponding ES starts to include non-negligible values for W_n with $n > 1$. This concept can be easily generalized for MCI projectiles, since the related PE involves ES with a clearly limited number of n (cf. Sect. 3.4). Scanning the projectile impact energy over the respective KE threshold region will reveal a typical change of the ES pattern from which the threshold value can be derived rather precisely.

This procedure should also be applicable with contaminated target surfaces where the PE is usually less important than for corresponding atomically clean targets.

4.2 On the Interdependence of PE and KE

In some earlier work on KE produced by impact of particles with keV energies no significant projectile charge state dependence of the total electron emission has been found (see e.g. Schram et al. 1966, Perdrix et al. 1968), whereas from more recent studies such a dependence became quite obvious, see e.g. Figs. 4.1a,b,c from Lakits and Winter (1990). This disagreement can at least partially be explained with the not atomically clean metal targets used in the earlier experiments, for which contributions from PE are generally strongly discriminated against those from KE.

It has already been mentioned that for impact of MCI the total electron yield passes with increasing impact velocity through a characteristic minimum, because of falling PE- and increasing KE contributions (cf. Sect. 3.2, in particular Figs. 3.9 and 3.10). If now the KE contribution is assumed as independent of the initial projectile charge and known from measurements for neutral or singly charged projectiles, it might be subtracted from apparent electron yields measured with higher q, to extract corresponding "exclusive" PE contributions. Such a procedure has been demonstrated in Figs. 4.2 and 4.3a (Fehringer et al. 1987). In addition, Fig. 4.3b demonstrates how well the assumption of a linear relationship between γ_{app} (the "apparent" total electron yield) and W_q (the total potential energy carried by a q-fold MCI) still holds in the presence of KE and in fact might be utilized for an approximate evaluation of the latters contribution. However, as already explained in Sect. 3.4 in some detail, from more recent experience gained in ES measurements on MCI-induced electron emission it became quite clear, that the mutual independence of PE and KE has to break down toward higher impact velocity. As soon as the projectile will arrive at the target surface in a still not fully neutralized state, it produces a more efficient KE contribution than an originally neutral or already completely neutralized particle of the same species and impact velocity. In Sect. 4.3 the reason for this difference will be further discussed.

To conclude the present considerations we may state, that only for sufficiently low ion impact velocities the contributions from PE and KE are mutually independent and can thus be disentangled by ways already outlined

---

Fig. 4.1a-c. Total electron yields derived from ES measurements for impact of respectively He^{q+} ($q = 0 \div 2$), Ne^{q+} ($q = 0 \div 3$) and Ar^{q+} ($q = 0 \div 4$) on clean gold, vs. impact velocity (Lakits and Winter 1990)

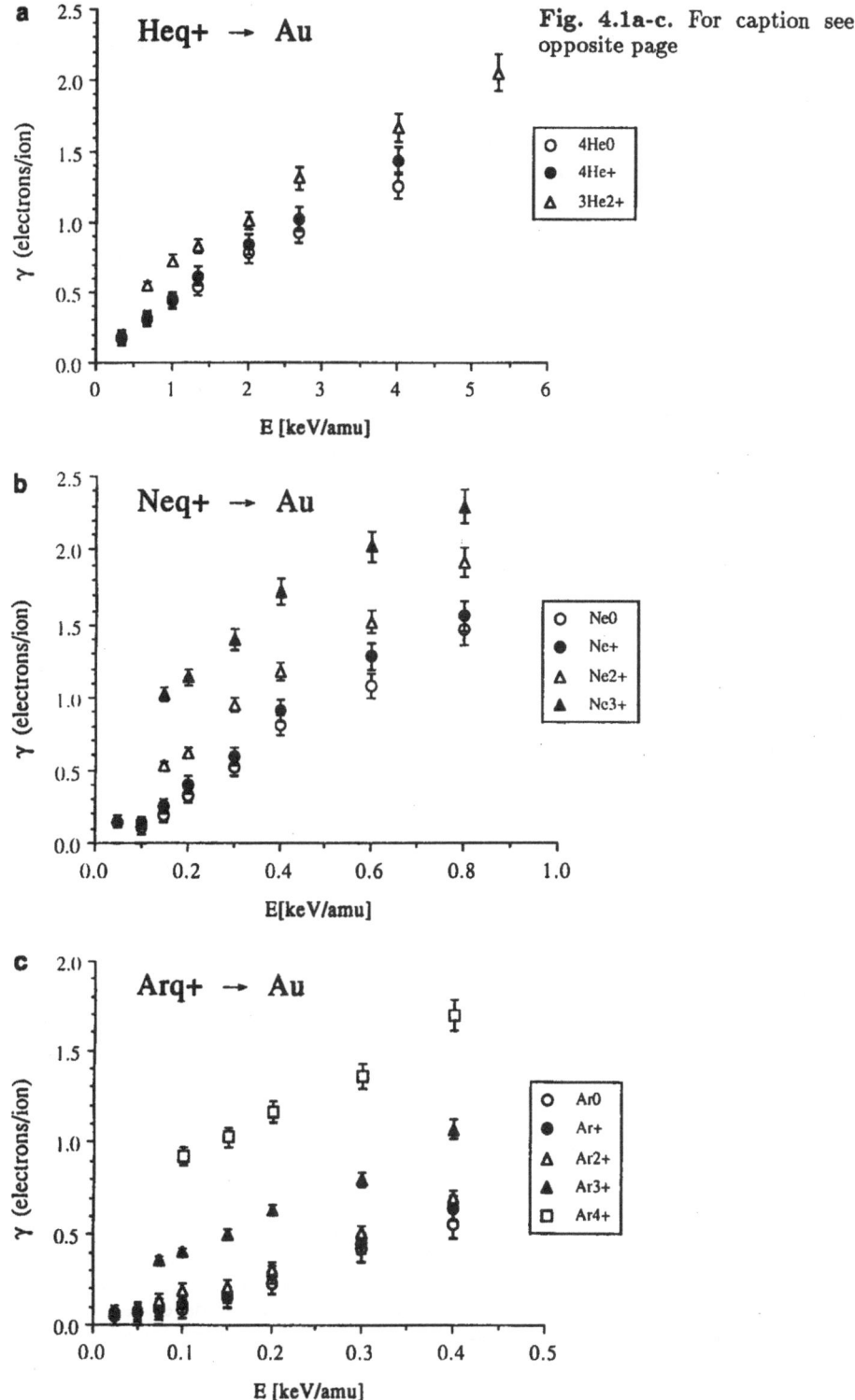

Fig. 4.1a-c. For caption see opposite page

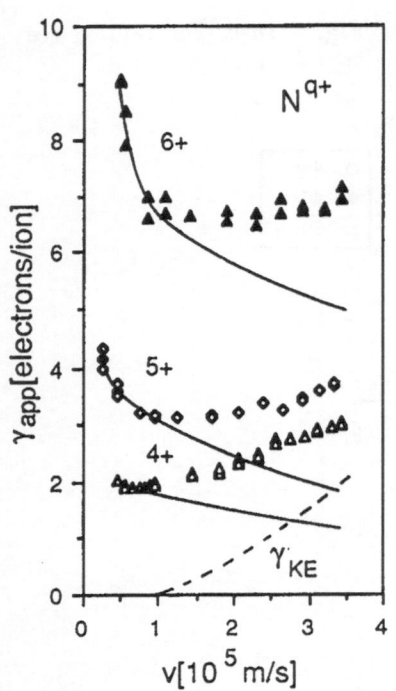

Fig. 4.2. Apparent total electron emission yields (symbols) and PE yields obtained by correction of the former data for the (assumed q-independent) kinetic emission yields (solid curves), vs. ion impact velocity, for impact of N^{q+} ($q = 4 \div 6$) on clean tungsten (Fehringer et al. 1987)

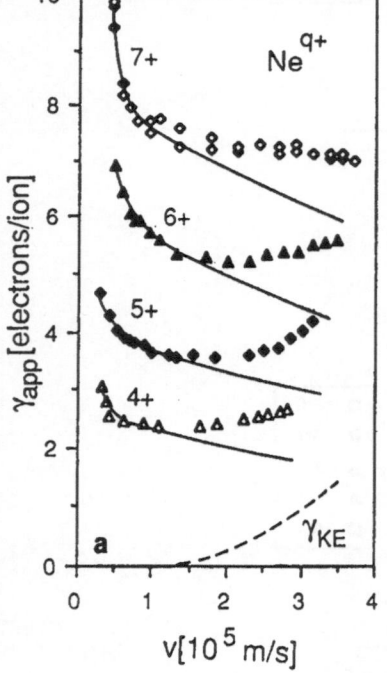

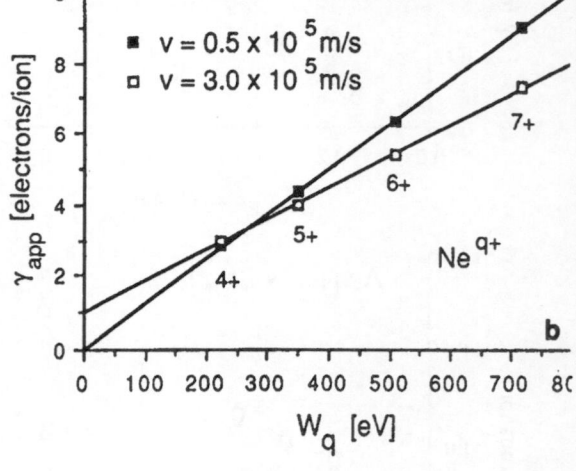

Fig. 4.3. (a) As for Fig. 4.2, but for impact of Ne^{q+} ($q = 4 \div 7$) on clean tungsten; (b) Course of apparent electron yield vs. total available potential energy, for impact of Ne^{q+} ($q = 4 \div 7$) on clean tungsten at two different impact velocities (Fehringer et al. 1987)

in Sect. 3.4. For higher impact energy the KE process will be initiated by projectiles which have not completely "forgotten" their initial charge state, and the distinction between PE and KE will loose its physical basis. We may still speak of PE and KE, but should rather base this distinction on the principal difference between electron emission initiated by projectiles still "outside" or already "inside" the target bulk, respectively.

4.3 Effects of Projectile Shielding in Slow Particle-Induced Kinetic Emission

Projectiles of given species and kinetic energy can be expected to interact in a slightly different fashion with solids if they carry with them a different number of electrons. This should be observable, e.g. for singly ionized ions in comparison with neutral projectiles, if the particles actually reach the surface in different charge state without undergoing resonance neutralization or -ionization processes (see previous section). In Fig. 4.5 we compare total electron yields for H^+-, H^0- and H^- impact on clean gold at impact energies between 1 and 16 keV, with sets of the relevant ES shown in Fig. 4.4 (Lakits et al. 1989b). A 1 keV proton is already so fast that within the distance where Auger transitions are probable (some 10^{-10} m), only a time of less than 10^{-14} s remains until surface impact, which is not sufficient to assure

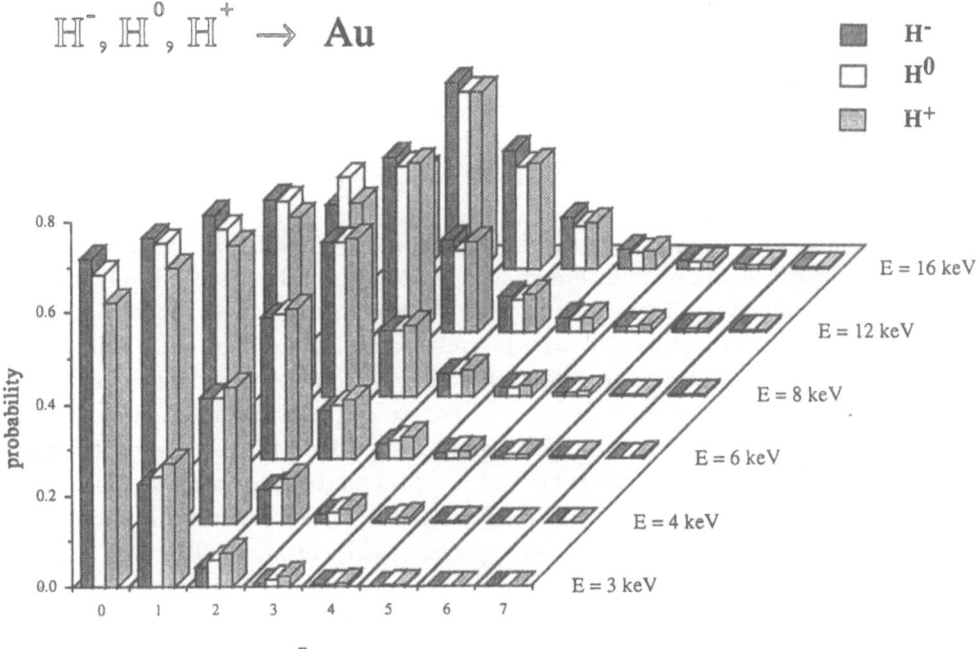

Fig. 4.4. ES for impact of respectively H^+, H^0 and H^- on clean gold, vs. impact energy (Lakits 1990)

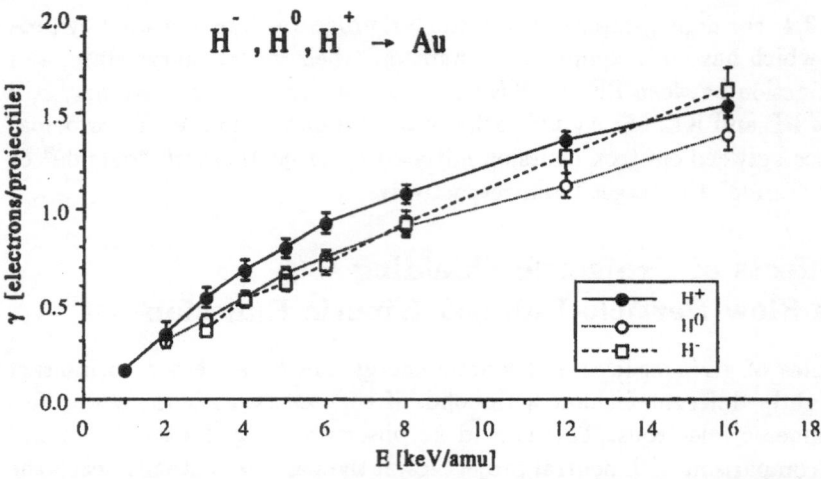

Fig. 4.5. Total electron yields derived from data shown in Fig. 4.4 (Lakits et al. 1989b)

complete neutralization. Consequently, a clearly measurable difference in total yields for impact of H^+ and H^0 is observed. Furthermore, for impact of H^- the total electron yield remains apparently the same as for H^0 up to ca. 8 keV, but then increases faster with E. This can be understood if we assume splitting off for one of the very loosely bound electrons from H^- upon its surface impact, after which the then free electron will contribute independently to the KE yield in the same way as an equally fast, but originally free electron in a "true" secondary electron emission process. The measured results shown in Fig. 4.5 could be well reproduced by calculations taking into account the charge-changing collisions for projectiles penetrating into the solid and the related electron production inside the solid. The latter process involves a penetration depth-dependent electronic stopping power due to a continuous change of the electron shielding.

As has been shown in detail by Lakits et al. (1990b), for a clean gold target the projectile equilibrium charge inside the solid is reached only after several atomic layers have been passed. Within this region the local projectile charge state remains still strongly related to the original one. This also concerns the resulting kinetic electron emission, since a considerable part of the electrons is being produced in the course of projectile interaction with the solid in the uppermost target region, from where they can escape into free space. Figure 4.6 shows results of such calculations in comparison with experimentally determined total electron yields for several combinations of neutral and singly ionized projectile species (Lakits et al. 1990b).

Effects due to "electronic shielding" of the projectile inside the target bulk are also responsible for the so-called "molecular effects" in kinetic emission, as can be demonstrated by comparing the impact of, respectively, atomic and molecular hydrogen ions on clean gold (Lakits 1990). Both the yields (by sum-

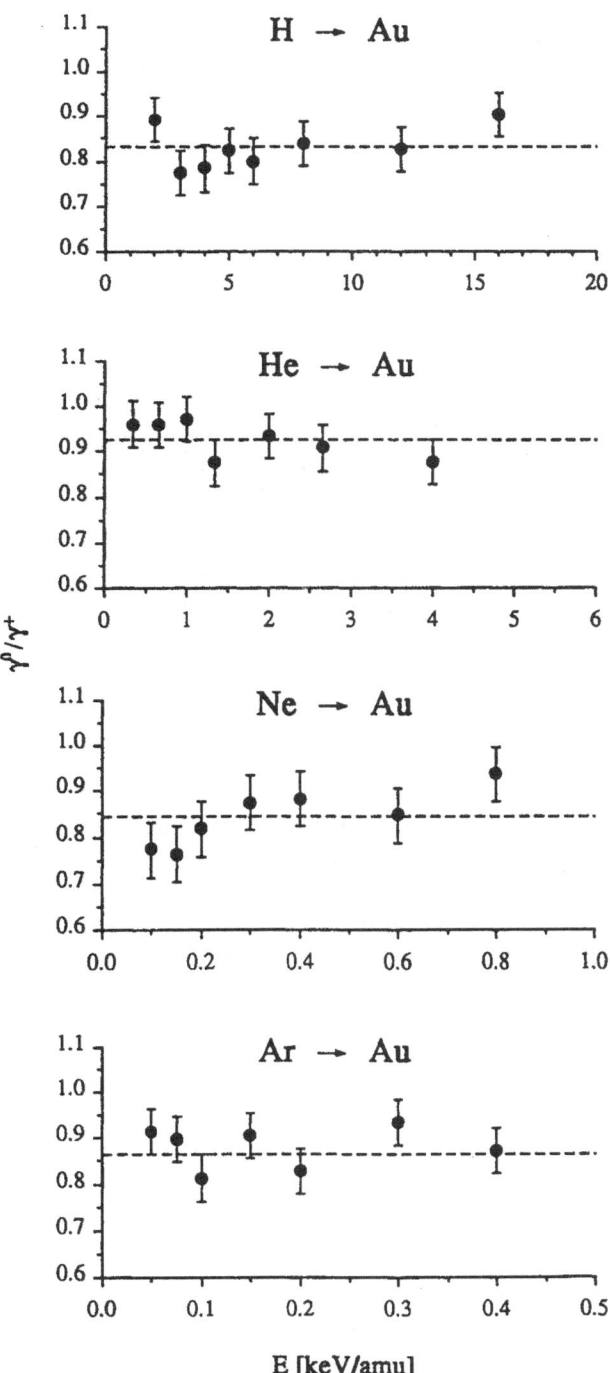

Fig. 4.6. Ratios of electron emission yields for impact of, respectively, neutral atoms and singly charged ions from H, He, Ne and Ar on clean gold, vs. impact velocity. Dashed lines show predictions as derived from model calculations (Lakits et al. 1990b)

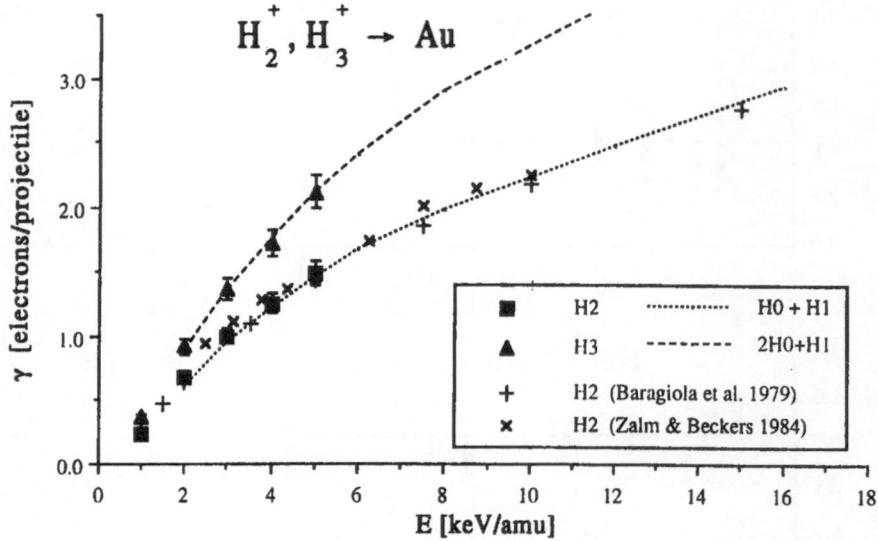

Fig. 4.7. Total electron yields vs. impact velocity for impact of singly charged hydrogen dimers and trimers, respectively on clean gold (symbols), in comparison with data from other authors. The curves show summed up data from yields obtained with equally fast neutral and singly ionized hydrogen atoms, as indicated (Lakits 1990)

mation, cf. Fig. 4.7) and the ES (by convolution, cf. Fig. 4.8a,b) for molecular ions can be synthesized from corresponding ES measurements for the equally fast atomic projectile constituents, if the correct number of accompanying electrons is taken into account. Measured ES can only be correctly reproduced by convolution of respectively one ES for H^+ and one ES for H^0 in case of a hydrogen dimer ion (Fig. 4.8a), and of respectively two ES for H^0 and one ES for H^+ in case of a hydrogen trimer ion (Fig. 4.8b). If for these convolutions exclusively proton-related ES are used, the synthesized results clearly differ from the actually observed ones.

Fig. 4.8. (a) ES for impact of singly ionized hydrogen dimers on clean gold, vs. impact velocity. Dark hatched barrows – directly measured ES; white barrows – ES convoluted from three measured ES for equally fast protons; light hatched barrows – ES convoluted from two measured ES for equally fast neutral hydrogen atoms and one measured ES for equally fast protons (Lakits et al. 1989b). (b) As for a), but for impact of singly ionized hydrogen trimers; picture assembled from data of Lakits (1990)

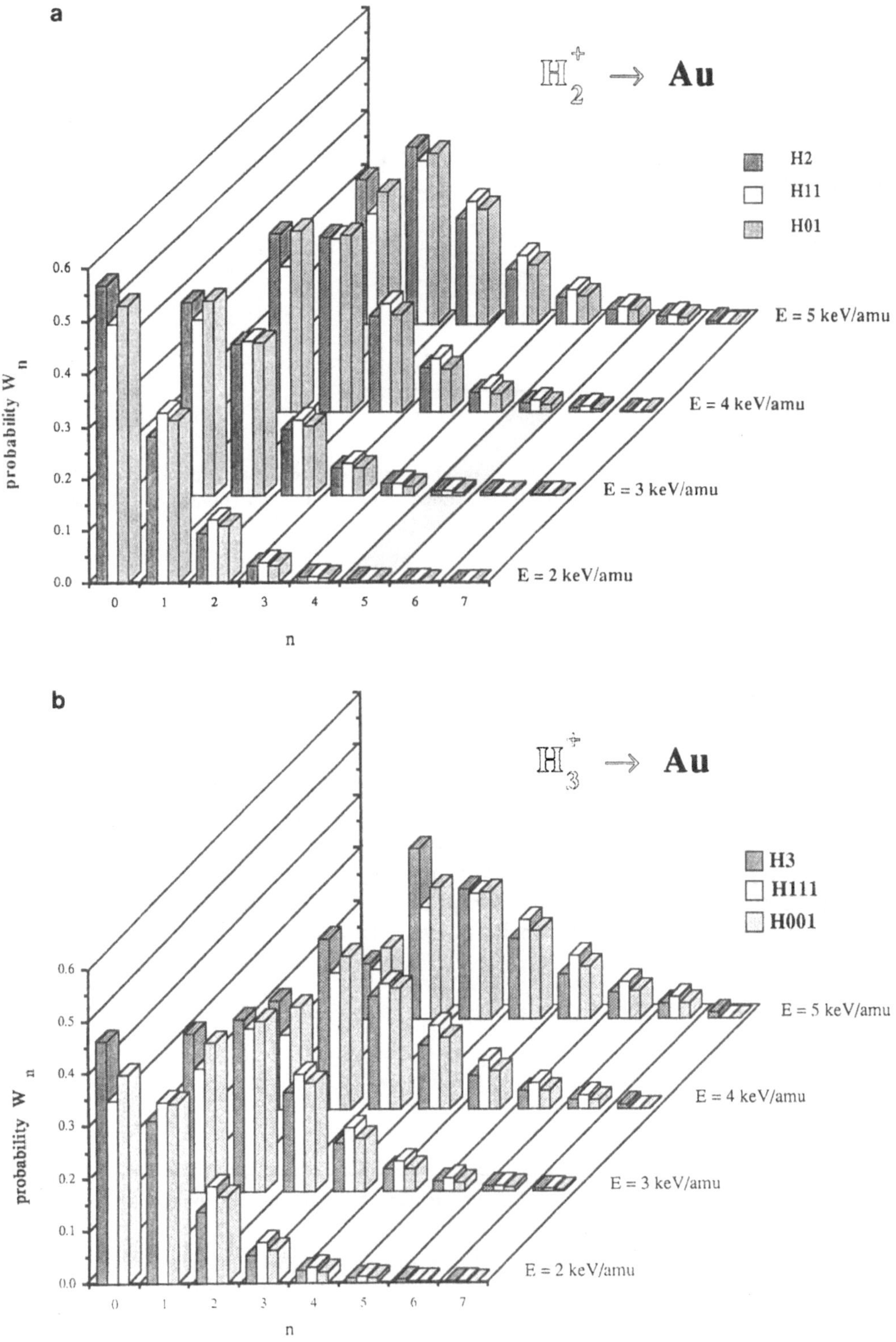
a
$H_2^+ \rightarrow$ Au
H2
H11
H01
probability W_n
0.6
0.5
0.4
0.3
0.2
0.1
0.0
0
1
2
3
4
5
6
7
n
E = 5 keV/amu
E = 4 keV/amu
E = 3 keV/amu
E = 2 keV/amu
b
$H_3^+ \rightarrow$ Au
H3
H111
H001
probability W_n
0.6
0.5
0.4
0.3
0.2
0.1
0.0
0
1
2
3
4
5
6
7
n
E = 5 keV/amu
E = 4 keV/amu
E = 3 keV/amu
E = 2 keV/amu

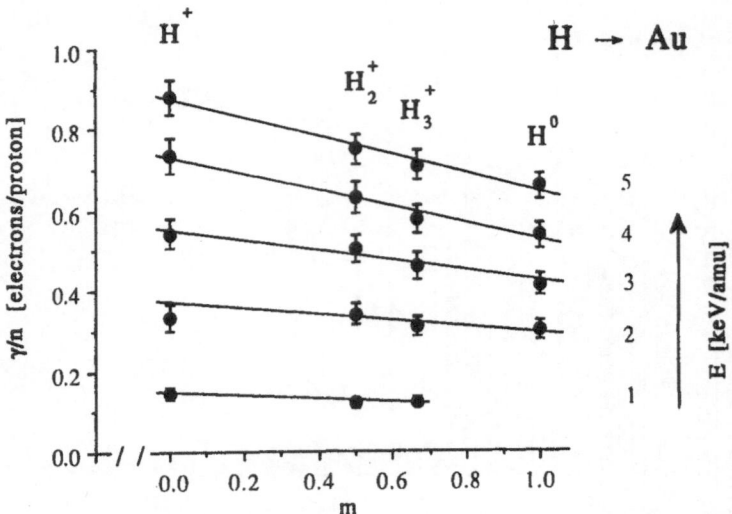

Fig. 4.9. Electron yield per proton vs. number m of electrons per proton carried by equally fast neutral and ionized hydrogen atoms, singly charged hydrogen dimers and trimers, respectively, toward a clean gold surface (Lakits et al. 1989b)

Figure 4.9, in a somewhat different way, demonstrates the same conclusion, i.e. the observed yields clearly depend on the number of shielding electrons per proton contained in the particular hydrogenic projectile (Lakits et al. 1989b).

Recently, such investigations have been extended to considerably lower impact energy (Winter et al. 1991). With ES data shown in Figs. 4.10a,b (note that for all three projectile species the PE is negligibly small), total yields for singly ionized hydrogen trimers could be well reconstructed both by summing up the "partial" yields and convolving the "partial" ES for two corresponding hydrogen dimer ions and one proton, as demonstrated in Fig. 4.11. Apparently, the above described concept of electronic shielding remains valid down to a projectile impact velocity of less than 5×10^4 m/s.

Fig. 4.10a,b. As for Figs. 3.24a,b, but for impact of singly charged hydrogen atoms, dimers and trimers, respectively (Winter et al. 1991)

Fig. 4.11. Total electron yields (symbols show measured values and solid line gives results obtained by convolution as for Fig. 4.8b) for impact of singly ionized hydrogen trimers on clean gold (Winter et al. 1991)

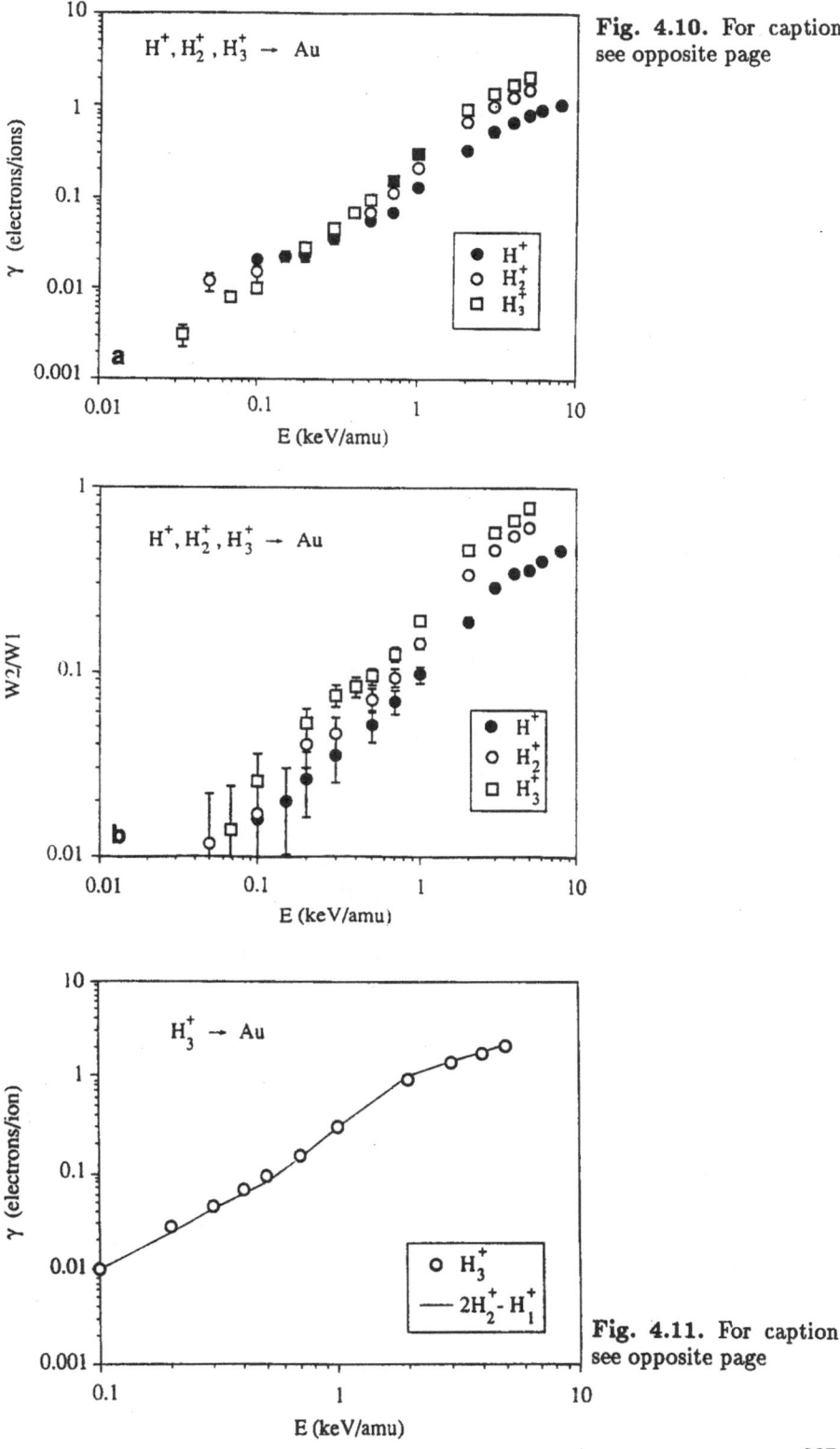

Fig. 4.10. For caption see opposite page

Fig. 4.11. For caption see opposite page

5. Conclusions and Outlook

We have reviewed the essential features of electron emission induced by impact of particles (ionized and/or highly excited atoms or molecules) on solid surfaces at such low impact velocities that kinetic effects [(i.e. electron emission produced as a result of kinetic energy transfer from the projectile onto target electrons – so-called kinetic emission (KE)] stay generally less important than electron emission due to the projectiles potential energy [so-called potential electron emission (PE)].

During neutralisation/deexcitation at the surface this potential energy can be imposed via various types of one-, two- ore many-electron transitions onto electrons to be extracted from the solid and emitted by way of the projectile.

We have described the principal experimental techniques for investigating PE and reviewed – in a rather simpleminded way – its theoretical background. Typical experimental results for total electron yields, ejected-electron energy distributions and electron emission statistics related to PE induced by singly, doubly or multiply charged projectiles from clean metal surfaces have been presented and discussed, and also the influence of target-surface adsorbates has been demonstrated by way of examples.

Neutralisation of highly charged projectiles approaching the surface can transiently generate projectile inner shell vacancies which decay under emission of fast Auger electrons and/or energetic photons. For a slow projectile this Auger electron emission can take place still before hitting the surface, whereas for faster projectiles the produced vacancies may either be transferred in close collisions to target atoms or decay after the projectile has been reflected from the surface. Consequently these fast Auger electrons constitute a "clock" for the time scale relevant for the PE transitions.

Informations of a similar kind are made available from comparison of electron emission statistics for equally fast projectiles in different charge states. In particular, the time can be determined within which full neutralisation of the projectiles is still achieved before hitting the surface. To use such "clocks" in a more quantitative way, further systematical investigations of PE processes for various projectiles and target surfaces would be highly desirable.

With increase of the impact velocity also KE will contribute to the total electron emission. Its share can be quantitatively determined as long as a full neutralisation before surface impact remains possible. Only in these cases it is physically meaningful to stick to the common distinction between PE and KE, whereas otherwise the instantaneous projectile charge state upon surface impact will considerably influence the subsequent KE. In the latter case it is more appropriate to speak of electron emission induced respectively prior to and after surface impact, instead of PE and KE. In addition, systematic studies of the angular dependence of PE are needed, and the recently observed finite response time of surfaces to electron extraction by approaching ions should also be studied in more detail.

In recent years the application of highly charged slow ion beams has provided gradually a much more detailed picture of the PE process and its close relations to the fields of MCI-induced photon and secondary ion emission, and MCI – induced desorption. Furthermore, emission of large specks of target matter ("Coulomb explosion") is expected to result from slow MCI impact on surface adsorbates and insulators, for which hopefully a clearcut experimental verification can soon be obtained.

Acknowledgements. Most experimental data presented in this review have been acquired in course of the theses of our former students Dr. Wilhelm Hofer (1983), Dr. Michael Fehringer (1987) and Dr. Gerhard Lakits (1990), which have been supported by Austrian "Fonds zur Förderung der Wissenschaftlichen Forschung" under project Nrs. 3283, S 18/04, 5317, S 43/04 and 6381.

We are grateful for collaboration and numerous contacts with other persons and groups to whose results we have referred at various occasions.

References

Andrä, H.J., 1989: in *Proceedings of NATO Summer School on Atomic Physics of Highly Charged Ions*, ed. R. Marrus, Plenum Press, and Nucl.Instr. and Meth. in Phys. Res. B **43**, 306

Alonso, E.V., R.A. Baragiola, J. Fèrron, M.M. Jakas and A. Oliva-Florio 1980: Phys. Rev. B **22**, 80

Alonso, E.V., M.A. Alurralde and R.A. Baragiola 1986: Surface Sci. **166**, L155

Amos A.T., K.W. Sulston and S.G. Davison 1989: in *Advances in Chemical Physics Vol 76, Molecule Surface Interactions*, ed. K.P. Lawley; John Wiley, p. 335

Anderson, P.W., 1961: Phys.Rev. **124**, 41

Apell, P. and R. Monreal, 1989: Nucl. Instrum. Meth. Phys. Res. B **42**, 171

Apell, P., 1987: Nucl. Instr. Meth. Phys. Res. B **23**, 242

Arifov, U.A., R.R. Rakhimov and K. Dzhurakulov 1962: Sov. Phys.-Doklady **7**, 209

Arifov, U.A., 1969: *Interaction of Atomic Particles with a Solid Surface*; Consultants Bureau, Plenum, New York

Arifov, U.A., 1971: Editor; *Secondary Emission and Structural Properties of Solids*; Consultants Bureau, Plenum, New York

Arifov, U.A., L.M. Kishinevskii, E.S. Mukhamadiev and E.S. Parilis, 1973: Zh. Tekh. Fiz. **43**, 181 (Sov. Phys.-Techn. Phys. **18** (1973) 118)

Arifov, T.U., E.K. Vasileva, D.D. Gurich, S.F. Kovalenko and S.N. Morozov, 1976: Izv. Akad. Nauk SSSR Ser. Fiz. **40**, 2621

Aumayr, F., G. Lakits and H. Winter 1991: Appl. Surface Sci. **47**, 139

Baragiola, R.A., E.V. Alonso, J. Fèrron and A. Oliva-Florio 1979: Surface Sci. **90**, 240

Bardsley, J.N., and B.M. Penetrante 1991: Proc. Symposium on Surface Science, Febr. 10–16, 1991, Obertraun/Austria, eds. P. Varga and G. Betz, T.U. Vienna, Austria, p. 3, and private communication

Bernhard, F., K.H. Krebs and I. Rotter 1965: Z. Physik **161**, 103

Bethe, H.E. and E.E. Salpeter, 1957: *Quantum Mechanics of One- and Two-Electron Systems*, Academic Press, New York

Bitensky, I.S., M.N. Murakhmetov and E.S. Parilis, 1979: Sov. Phys. Tech. Phys. **24**, 618

Bitensky, I.S. and E.S. Parilis, 1989: J. de Physique **C2**, 227

Blandin A., A. Nourtier and D.W. Howe, 1976: J. de Physique **37**, 369

Bloss W. and D. Hone 1978: Surf. Science **72**, 277

Boiziau, C., 1981: in *Inelastic Particle-Surface Collisions*, ed. by E. Taglauer and W. Heiland, Springer Ser. Chem.Phys. 17, Springer, p. 48

Brako, R., and D.M. Newns 1985a: Sol. St. Comm. **55**, 633

Brako, R., and D.M. Newns 1985b: Physica Scripta **32**, 451

Briand, J.P., L. de Billy, P. Charles, S. Essabaa, P. Briand, R. Geller, J.P. Desclaux, S. Bliman and C. Ristori, 1990: Phys. Rev. Letters **65**, 159

Brown, I.G., 1989: *The Physics and Technology of Ion Sources*, John Wiley, New York

Cano, G.L., 1973: J. Appl. Phys. **44**, 5293

Carter, G. and J.S. Colligon, 1968: Chapter 3 in *Ion Bombardment of Solids*, Heinemann, London

Chaban, E.E., H.D. Hagstrum and P. Petrie, 1990: Rev. Sci. Instr. **60**, 3647

Delaney, C.F.G. and P.W. Walton 1966: IEEE Trans. Nucl. Sci. **NS-13**, 742

Delaunay, M., S. Dousson, R. Geller, P. Varga, M. Fehringer and H. Winter, 1985: *Abstracts of Contributed Paper, XIV ICPEAC Palo Alto*, eds. M.J. Coggiola, D.L. Huestis, R.P. Saxon, p. 477

Delaunay, M., M. Fehringer, R. Geller, D. Hitz, P. Varga and H. Winter, 1987a: Phys. Rev. B **35**, 4232

Delaunay, M., M. Fehringer, R. Geller, P. Varga and H. Winter, 1987b: Europhys. Lett. **4**, 377

Donets, E.D., 1983: Physica Scripta **T3**, 11

Donets, E.D., S.V. Kartashov, V.P. Ovsyannikov, 1985: Proceedings 17. Intern. Conf. on Phenomena in Ionized Gases, Budapest, and private communication

Erickson, R.L. and D.P. Smith, 1975: Phys. Rev. Lett. **34**, 297

Ertl, G., 1986: Phil. Trans. R. Soc. Lond. **A 318**, 51

Fehringer, M. 1987: thesis, Technische Universität Wien, Austria (unpublished)

Fehringer, M., M. Delaunay, R. Geller, P. Varga and H. Winter 1978: Nucl. Instr. and Meth. Phys. Res. B **23**, 245

Folkerts, L. and R. Morgenstern 1990 Europhys. Letters **13**, 377

Gemmell, D.S., 1974: Rev. Modern Phys. **46**, 129

Gilbody, H.B., 1978: *Low Energy Ion Beams 1977*, Inst. Phys. Conf. Series No. 38, Bristol, ch. 4

Grunze, M., H. Ruppender and O. Elshazly, 1988: J. Vac. Sci. Technol. **A6**, 1266

Hentschke, R., K.J. Snowdon, P. Hertel and W. Heiland 1986: Surface Sci. **173**, 565

Hofer, W., 1983: thesis, Technische Universität Wien, Austria (unpublished)
Hofer, W., W. Vanek, P. Varga and H. Winter 1983a Surface Sci. **126**, 605
Hofer, W., W. Vanek, P. Varga and H. Winter 1983b Rev. Sci. Instrum. **54**, 150
Hofer, W. and P. Varga, 1984: Nucl. Instr. and Meth. Phys. Res. B **2**, 391
Hagstrum, H.D., 1953: Rev. Sci. Instrum. **24**, 1122
Hagstrum, H.D., 1954a: Phys. Rev. **96**, 325
Hagstrum, H.D., 1954b: Phys. Rev. **96**, 336
Hagstrum, H.D., 1956a: Phys. Rev. **104**, 672
Hagstrum, H.D., 1956b: Phys. Rev. **104**, 317
Hagstrum, H.D., 1956c: Phys. Rev. **104**, 1516
Hagstrum, H.D,. 1956d: Phys. Rev. **104**, 309
Hagstrum, H.D., 1956e: Phys. Rev. **104**, 672
Hagstrum, H.D., 1960: J. Appl. Phys. **31**, 897
Hagstrum, H.D., D.D. Pretzer and Y. Takeishi, 1965: Rev. Sci. Instrum. **36**, 1183
Hagstrum, H.D. and G.E. Becker, 1973: Phys. Rev. B **8**, 107
Hagstrum, H.D., 1977: in *Inelastic Ion-Surface Collisions*, ch 1, eds.: N.H. Tolk,
    J.C. Tully, W. Heiland and C.W. White, Academic press, New York
Hagstrum, H.D., 1978: in *Electron and Ion Spectroscopy of Solids*, eds.: L. Fiermans,
    J. Vennik, W. Dekeyser; Plenum Press, New York, p. 273
Hasselkamp, D., 1991: *Springer Tracts in Modern Physics*, this volume

Kaminsky, M., 1965: *Atomic and Ionic Impact Phenomena on Metal Surfaces*,
    Springer, Berlin chs. 12, 13
Kasai, H., and A. Okiji 1987: Surface Science **183**, 147
Kishinevskii, L.M. 1973: Radiation Effects **19**, 23
Köhrbrück, R., D. Lecler, F. Fremont, P. Roncin, K. Sommer, T.J.M. Zouros, J.
    Bleck-Neuhaus and N. Stolterfoht 1991: Nucl. Instrum. Meth. Phys. Res. B
    56, 219

Lakits, G., F. Aumayr and H. Winter 1989a: Rev. Sci. Instrum. **60**, 3151
Lakits, G., F. Aumayr and H. Winter 1989b: Europhys. Letters **10**, 679
Lakits, G. 1990: thesis, Technische Universität Wien, Austria (unpublished)
Lakits, G. and H. Winter 1990: Nucl. Instrum. Meth. Phys. Res. B **48**, 597
Lakits, G., F. Aumayr, M. Heim and H. Winter 1990a: Phys. Rev. A **42**, 5780
Lakits, G., A. Arnau and H. Winter 1990: Phys. Rev. B **42**, 15

McDaniel, E.W., 1964: *Collision Phenomena in Ionized Gases*, John Wiley, New
    York, chr. 13
Medved, D.B., P. Mahadevan and J.K. Layton 1963: Phys. Rev. **129**, 2086
Misković, Z.L. and R.K. Janev 1989: Surface Sci. **221**, 317
Musket, R.G., W. McLean, C.A. Colmenares, D.M. Makowiecki and W.J. Siekhaus,
    1982: Appl. of Surface Science **10**, 143

Newns D.M., 1969: Phys. Rev. **178**, 1123

Oda, K., A. Ichimiya, Y. Yamada, T. Yasue and S. Ohtani 1988 Nucl. Instrum.
    Meth. Phys. Res. B **33**, 345

Perdrix, M., S. Paletto, R. Goutte and C. Guillaud 1969: Phys. Letters **28A**, 534
Petrov, N.N. 1960: Fiz. Tverd. Tela **2**, 1300

Schram, B.L., A.J.H. Boerboom, W. Kleine and J. Kistemaker 1966: Proc. 7. Int. Conf. Phenomena in Ionized Gases, Beograd, p. 170

Schweinzer, J. and H. Winter 1989: J. Phys. B: At. Mol. Opt. Phys. **22**, 893

Scoles, G. (ed.), 1988: *Atomic and Molecular Beam Methods*, vol. 1, Oxford Univ. Press, Oxford

Sesselmann, W., B. Woratschek, J. Küppers and G. Ertl, 1987a: Phys. Rev. B **35**, 1547

Sesselmann, W., B. Woratschek, J. Küppers and G. Ertl, 1987b: Phys. Rev. B **35**, 8348

Shekhter, S.S., 1937: J. Exptl. Theor. Phys.(USSR.) **7**, 750

Snowdon, K.J., R. Hentschke, A. Närmann and W. Heiland, 1986: Surf. Sci. **173**, 581

Snowdon, K.J., 1988: Nucl.Instr. Meth. Phys. Res. B **34**, 309

Staudenmaier, G., W.O. Hofer and H. Liebl 1976: Int. J. Mass Spectrosc. Ion Phys. **11**, 103

Sulston, K.W., A.T. Amos and S. G. Davison 1988: Phys.Rev. B **37**, 9121

Taglauer E., 1990: Appl. Phys. A **51**, 238

Tully J.C., 1977: Phys. Rev. B **16**, 4324

Varga, P. 1978: thesis, Technische Universität, Wien, Austria (unpublished)

Varga, P., and H. Winter 1978: Phys. Rev. A **18**, 2453

Varga, P., W. Hofer and H. Winter 1981: J. Phys. B: At. Mol. Phys. **14**, 1341

Varga, P., W. Hofer and H. Winter, 1982a: J. Scanning Electron Microsc. **1**, 967

Varga, P., W. Hofer and H. Winter, 1982b: Surf. Sci. **117**, 142

Varga, P., 1987: Appl. Phys. A **44**, 31

Varga, P., 1988: in *Electronic and Atomic Collisions*, 793, eds. H.B. Gilbody, W.R. Newell, F.H. Read and A.C.H. Smith, North Holland, Amsterdam

Varga, P., 1989: Comm. At. Mol. Phys. **23**, 11

Waters, P.M. 1958: Phys. Rev. **111**, 1053

Winter, H., and R. Zimny, 1988: in *Coherence in Atomic Collision Physics*, eds. H.J. Beyer, K. Blum and R. Hippler, Plenum, New York, p. 283

Winter, H. 1991: Z. Phys. D: At. Mol. Clusters **21**, S129

Winter, H., F. Aumayr and G. Lakits 1991: Nucl. Instrum. Meth. Phys. Res. B (in print)

Woratschek B., W. Sesselmann, J. Küppers, G. Ertl and H. Haberland 1985a: Phys. Rev. Lett. **55**, 611

Woratschek B., W. Sesselmann, J. Küppers and G. Ertl 1985b: Phys. Rev. Lett. **55**, 1231

Wouters, P.A.A.F., P.A. Zeijlmans van Emmichoven and A. Niehaus 1989 Surface Sci. **211/212**, 249

van Wunnik, J.N.M., J.J.C. Geerlings and J. Los, 1983: Surf. Sci. **131**, 1

Zalm, P.C., and L.J. Beckers 1984: Philips J. Res. **39**, 61

Zehner, D.M., S.H. Overbury, C.C. Havener, F.W. Meyer and W. Heiland, 1986: Surf. Sci. **178**, 359

Zeijlmans van Emmichoven, P.A., P.A.A.F. Wouters and A. Niehaus, 1988: Surf. Sci. **195**, 115

Zeijlmans van Emmichoven, P.A. and A. Niehaus 1990: Comm. At. Mol. Phys. **24**, 65

de Zwart, S.T., T. Fried, U. Jellen, A.L. Boers and A.G. Drentje 1985: J. Phys. B: At. Mol. Phys. **18**, L623

de Zwart, S. T., 1987: Thesis, University of Groningen, The Netherlands (unpublished)

de Zwart, S. T., A.G. Drentje, A.L. Boers and R. Morgenstern, 1989: Surface Sci. **217**, 298

List of Abbreviations

AD	Auger de-excitation
AES	Auger electron spectroscopy
AI	auto-ionization
AN	Auger neutralization
ES	electron emission statistics
FWHM	full width at half maximum
INS	Ion neutralization spectroscopy
ISS	Ion-surface scattering
KE	kinetic electron emission
KLL	Auger transition combining a L-K transition with ionization of a second L electron
LEED	Low energy electron diffraction
LMM	Auger transition combining a M-L transition with ionization of a second M electron
MCI	multicharged ion
MDS	metastable atom de-excitation spectroscopy
PE	potential electron emission
PES	photo-electron spectroscopy
PIE	particle-induced electron emission
QRN	quasi-resonance neutralization
RD	radiative de-excitation
RI	resonance ionization
RN	resonance neutralization
S-DOS	surface density-of-states
sPIE	slow particle-induced electron emission
UHV	ultra-high vacuum

Subject Index

N. G. Chetaev

Theoretical Mechanics

Translated from the Russian by I. Aleksanova

1989. 407 pp. 190 figs. Hardcover
ISBN 3-540-51379-5

This university-level textbook reflects the extensive teaching experience of N. G. Chataev, one of the most influential teachers of theoretical mechanics in the Soviet Union. The mathematically rigorous presentation largely follows the traditional approach, supplemented by material not covered in most other books on the subject. To stimulate active learning numerous carefully selected exercises are provided. Attention is drawn to historical pitfalls and errors that have led to physical misconceptions.

Extensive appendices contain material from additional lectures on optics and mechnics analogies, Poincaré's equation and the special theory of elasticity.

**Distribution rights for the socialist countries, India and Iran:
V/O "Mezhdunarodnaya Kniga", Moscow**

D. Park, Williams College, Williamstown, MA

Classical Dynamics and Its Quantum Analogues

2nd enl. and updated ed. 1990. IX, 333 pp. 101 figs. Hardcover ISBN 3-540-51398-1

The primary purpose of this textbook is to introduce students to the principles of classical dynamics of particles, rigid bodies, and continuous systems while showing their relevance to subjects of contemporary interest. Two of these subjects are quantum mechanics and general relativity. The book shows in many examples the relations between quantum and classical mechanics and uses classical methods to derive most of the observational tests of general relativity. A third area of current interest is in nonlinear systems, and there are discussions of instability and of the geometrical methods used to study chaotic behaviour. In the belief that it is most important at this stage of a student's education to develop clear conceptual understanding, the mathematics is for the most part kept rather simple and traditional.

This book devotes some space to important transitions in dynamics: the development of analytical methods in the 18th century and the invention of quantum mechanics.

A. Hasegawa, AT & T Bell Laboratories,
Murray Hill, NJ

Optical Solitons in Fibers

2nd enl. ed. 1990. XII, 79 pp. 25 figs.
Softcover ISBN 3-540-51747-2

Already after six months high demand made a new edition of this textbook necessary. The most recent developments associated with two topical and very important theoretical and practical subjects are combined: **Solitons** as analytical solutions of nonlinear partial differential equations and as lossless signals in dielectric **fibers.** The practical implications point towards technological advances allowing for an economic and undistorted propagation of signals revolutionizing telecommunications. Starting from an elementary level readily accessible to undergraduates, this pioneer in the field provides a clear and up-to-date exposition of the prominent aspects of the theoretical background and most recent experimental results in this new and rapidly evolving branch of science. This well-written book makes not just easy reading for the researcher but also for the interested physicist, mathematician, and engineer. It is well suited for undergraduate or graduate lecture courses.

Springer-Verlag
Berlin
Heidelberg
New York
London
Paris
Tokyo
Hong Kong
Barcelona
Budapest

A. G. Sitenko, Academy of the Ukrainian SSR

Scattering Theory

1991. XI, 294 pp. 32 figs. (Springer Series in Nuclear and Particle Physics) Hardcover ISBN 3-540-51953-X

This book is an introduction to nonrelativistic scattering theory. The presentation is mathematically rigorous, but is accessible to upper level undergraduates in physics. The relationship between the scattering matrix and physical observables, i.e. transition probabilities, is discussed in detail. Among the emphasized topics are the stationary formulation of the scattering problem, the inverse scattering problem, dispersion relations, three-particle bound states and their scattering, collisions of particles with spin and polarization phenomena. The analytical properties of the scattering matrix are discussed. Problems round off this volume.

B. N. Zakhariev, Moscow; A. A. Suzko, Minsk, USSR

Direct and Inverse Problems

Potentials in Quantum Scattering

1990. XIII, 223 pp. 42 figs. Softcover ISBN 3-540-52484-3

This textbook can almost be viewed as a "how-to" manual for solving quantum inverse problems, that is, for deriving the potential from spectra or scattering data and also, as somewhat of a quantum "picture book" which should enhance the reader's quantum intuition. The formal exposition of inverse methods is paralleled by a discussion of the direct problem. Differential and finite-difference equations are presented side by side. The common features and (dis)advantages of a variety of solution methods are analyzed. To foster a better understanding, the physical meaning of the mathematical quantities are discussed explicitly. Wave confinement in continuum bound states, resonance and collective tunneling, energy shifts and the spectral and phase equivalence of various interactions are some of the physical problems covered.

K. L. G. Heyde, University of Gent, Belgium

The Nuclear Shell Model

1990. XII, 376 pp. 171 figs. (Springer Series in Nuclear and Particle Physics) Hardcover ISBN 3-540-51581-X

This book evolved from a course in theoretical nuclear physics taught over seven years at the University of Gent and is thus well suited to and tested for lecture courses. The nuclear shell model is introduced from basic techniques such as angular momentum and tensor algebra. The material is developed from the beginning up to the present state-of-the-art calculations using self-consistent residual interactions. Problem sets and simple computer codes are included to facilitate a better acquaintance with the subject.
The appendices constitute an integral part of the text going into depth on a number of technical derivations to provide the reader with a detailed background facilitating active research. The book introduces the subject to advanced undergraduate and to graduate students providing them with knowledge and techniques for own research in this field. It is a highly useful prerequisite for lecturers teaching modern nuclear physics.

Springer-Verlag
Berlin
Heidelberg
New York
London
Paris
Tokyo
Hong Kong
Barcelona
Budapest

Springer Tracts in Modern Physics

* denotes a volume which contains a Classified Index starting from Volume 36